Dynamics of Agricultural Extension Approaches

Dynamics of Agricultural Extension Approaches

— Editors —

A.K. Singh

Assistant Director General (Agril. Extension)
Indian Council of Agricultural Research
Krishi Anusandhan Bhawan, Pusa, New Delhi – 110 002

Jitendra Chauhan

Head, Department of Agricultural Extension
R.B.S. College, Bichpuri, Agra, Uttar Pradesh

R. Roy Burman

Senior Scientist, Division of Agricultural Extension
Indian Agricultural Research Institute, New Delhi – 110 002

K. Sandhu

Assistant Professor, (Home Science)
Punjab Agriculture University, Ludhiana, Punjab

BIOTECH

2015

BIOTECH BOOKS®

ISBN 978-81-7622-342-3

Published by: **BIOTECH BOOKS®**
4762-63/23, Ansari Road, Darya Ganj,
New Delhi - 110 002
Phone: +91-011-23262132
E-mail: biotechbooks@yahoo.co.in

Printed at: **Chawla Offset Printers**
Delhi - 110 052

PRINTED IN INDIA

Preface

Agricultural extension plays a fundamental role in the sustainable agricultural development. To cater the needs of farmers, agricultural knowledge network is widening in a big way with the emergence of pluralistic extension partners. Learning is a continuous process and it is becoming more dynamic with multiplicity of domains. Thus, there is always a need to understand different concepts, techniques and methods which help in development of strategic planning, effective implementation, professional management and systematic evaluation of an extension programme. The task and responsibility of extension education is need based and holistic in content and scope, thus beyond agricultural technology transfer. It is very important that knowledge on agricultural extension be constantly improved, rationalized, and adapted.

The Society of Extension Education, Agra, Uttar Pradesh has been organizing National Extension Education Congress on different issues and themes in order to sensitize the agricultural extension professionals about various ideologies, concepts, processes and models being evolved during the process of pursuing different programmes. During the course of such interfaces, a number of useful contributions were received from very experienced and knowledgeable scholars. The Executive of the Society viewed to select some important contributions and bring out a publication namely "Dynamics of Agricultural Extension Approaches" to enlighten extension professionals in the rapidly changing scenario and potential role of extension in this process.

We believe that this publication will be useful for students, teachers, trainers, educators, field practitioners and managers. We owe to express our gratitude to all the contributors.

Editors

Contents

List of Contributors

Bara, Niva
Head, Deptt. of Agril. Extension, BAU, Ranchi-834006 (Jharkhand)

Burman, R. Roy
Sr. Scientist, Division of Agriculture Extension, IARI, New Delhi-110012

Chauhan, Jitendra
Joint Director (KVKs) and Head, Agriculture Extension, RBS College, Bichpuri, Agra – 283 105

Das, P.
Former Deputy Director General, Agricultural Extension, ICAR, New Delhi–110012

Deshmukh, B.A.
Junior Research Assistant, Mahatma Phule Krishi Vidyapeeth, Rahuri (M.S.)

Dubey, S.K.
Sr. Scientist (Agriculture Extension), Zonal Project Directorate (ICAR), Zone IV, Kanpur, UP

Gautam, U.S.
Principal Scientist, Zonal Project Director, Zone VII, JNKVV, Jabalpur

Ghosh, Souvik
Professor (Agril. Extension), Deptt. of Agril. Extension, Agril. Economics and Agril. Statistics, Institute of Agriculture, Visva-Bharati University, Sriniketan-731236, West Bengal

Gupta, H.S.
Former Director , IARI, New Delhi-110012

Gupta, Vinod
SMS, KVK, Jammu, Sher-e-Kashmir University of Agricultural Sciences & Technology-Jammu

Hansra. B.S.
Former ADG (AE) ICAR and Presently Director, School of Agriculture, IGNOU, New Delhi-110068

Jain, P.K.
Assistant Professor, School of Agriculture, IGNOU, New Delhi-110068

Kharde, P.B.
Associate Professor of Agril. Extn., Mahatma Phule Krishi Vidyapeeth, Rahuri (M.S.)

Kokate, K.D.
Former ADG (AE) ICAR and Presently Director, , Extension, Mahatma Phule Krishi Vidyapeeth, Rahuri (M.S.)

Kumar, Ashwani
Director, Directorate of Water Management (ICAR), Bhubaneswar

Kumar, S. Prabhu
Zonal Coordinator, Zonal Project Directorate, Zone I, Ludhiana

Mohapatra, Tanmaya
Senior Research Fellow, Directorate of Water Management (ICAR), Bhubaneswar

Mondal. Biswajit
Scientist (Ag. Econ.), Central Soil and water Conservation Research and Training Institute (CSWCRTI), Research Centre, Bellary – 583104 (Karnataka)

Narayana Gowda, K.
Former Vice-chancellor, University of Agricultural Sciences, Bangalore, Karnatak

Narayana Swamy, B.K.
Professor of Extension and Head, Department of Distance Education, College of Agriculture (North Block), GKVK, Bangalore-560065

Pal, P.P.
Sr. Scientist, Zonal Project Directorate, Zone-II, Salt Lake, Kolkata

Patil, S.S.
Assistant Professor of Agril. Extn., Mahatma Phule Krishi Vidyapeeth, Rahuri

Prasad, C.
Former Deputy Director General, Agricultural Extension, ICAR, New Delhi–110012

Rai, Pradeep Kumar
SMS, KVK, Jammu, Sher-e-Kashmir University of Agricultural Sciences & Technology-Jammu

Risam, K.S.
Director, Directorate of Extension Education, Sher-e-Kashmir University of Agricultural Sciences & Technology-Jammu

Sairam, C.V.
Principal Scientist (Agrl. Econ.) of Zonal Project Directorate, Bangalore

Sarkar, A.
Joint Director of Extension Education, B.C.K.V., Mohanpur, West Bengal

Sekar, I.
Principal Scientist (Ag. Econ.), Division of Agricultural Economics, Indian Agricultural Research Institute (IARI), Pusa, New Delhi – 110 012, India

Shanthy, T. Rajula
Principal Scientist and Head, Extension Section, Sugarcane Breeding Institute (ICAR), Coimbatore – 641 007, Tamil Nadu

Sharma, J.P.
Joint Director (Extension), IARI, New Delhi-110012

Singh, A.K.
Former Deputy Director General, Agricultural Extension, ICAR, New Delhi–110012

Singh, A.K.
Zonal Project Director, Zonal Project Directorate, Zone-II, Salt Lake, Kolkata-700097 (W.B.)

Singh, A.K.[1]
Zonal Project Director, Zone-IV, Kanpur-208002 (U.P.)

Singh, Alka
SMS (HSc.), KVK, Sidhi, JNKVV, Jabalpur

Singh, Alka
Principal Scientist (Ag. Econ.), Division of Agricultural Economics, Indian Agricultural Research Institute (IARI), Pusa, New Delhi – 110 012, India

Singh, Baldeo
Former Joint Director (Extension), IARI, New Delhi-110012

Singh, Daulat
Former Director Extension, CSAUAT, Kanpur-208 002

Singh, Lakhan
Principal Scientist (Agriculture Extension), Zonal Project Directorate (ICAR), Zone IV, Kanpur-208 002, U.P.

Singh, R.P.[1]
Former Vice-chancellor, MPUA&T, Udaipur, Rajasthan

Singh, R.P.
Associate Director Extension, G B Pant University of Agri. & Tech. Pantnagar (Uttarakhand)

Singh, R.P. 'Ratan'
Director Extension Education, BAU, Ranchi-834006 (Jharkhand)

Singh, S.R.K.
Senior Scientist (AE), ZPD Zone VII, Jabalpur

Sinha, B.P.
Former Head, Division of Agriculture Extension, IARI, New Delhi-110012

Srivastava, R.C.
Former, Director, Central Agricultural Research Institute, Port Blair – 744 101

Veerabhadraiah, V.
Former Director of Extension, University of Agricultural Sciences, Bangalore, Karnatak

Vijayaragavan, K.
Former Joint Director (Extension), IARI, New Delhi-110012

Chapter 1

Emerging Models of Technology Application for Agri-Rural Development

R.P. Singh[1]

The Indian economy has its mainstay on agriculture and allied activities. Food is basic to life and wellbeing and has always been taken on the highest of all our priorities. Agriculture in India has traveled a long distance – Green revolution, White revolution, Yellow revolution, Blue revolution, Rainbow revolution to thrust on Ever Green Revolution – the stability and sustainability that we have to attain.

Science and technology play a vital role in the development strategy of various section of the society. Technological empowerment and sustainable livelihood at the grass root level is very much essential at this juncture of our development. The intervention of the developments in Science and technology into the field of agriculture, industry, education and health is very much crucial particularly in rural situation.

In the perspective of modern scientific intervention in the present social scenario, this 6th National Extension Education Congress on Emerging models of technology application for agri-rural development would deliberate on the following five important theme areas.

Climate Change: Ground Realities

Climate change and agriculture are interrelated, both of which influence each other on a global scale. Global warming due to changes temperature, precipitation and carbon dioxide is projected to have significant impacts on conditions affecting agriculture, glacial run-off, water resources and the interaction of these elements. These conditions determine the carrying capacity of the biosphere to produce enough food for the human population and domesticated animals. The overall effect of climate change on agriculture will depend on the balance of these effects.

The vulnerability of agriculture to climate change is well acknowledged. But what is not fully appreciated is the impact this will have on Indian agriculture, more so in rain-fed (non-irrigated), practiced mostly by small and marginal farmers who will suffer the most. The crops that may be hit include, cereals, pulses and oilseeds, among others pulses, oilseeds are already in short supply and are consequently high-priced.

Nearly 80 million hectares, out of the country's net sown area of around 143 million hectares, lack irrigation facilities and, hence, rely wholly on rain water for crop growth. Over 85 per cent of the pulses and coarse cereals, more the 75 per cent of the oilseeds and nearly 65 per cent of cotton are produced from such lands. The crop yields are quite low. Most of the rain-fed lands, moreover, are in arid and semi-arid zones where annual rainfall is meager and prolonged dry spells are quite usual even during the monsoon season. This makes crop cultivation highly risk-prone. If the quantum of rainfall in these areas drops further or its pattern undergoes any distinct, albeit unforeseeable, change in the coming years, which seems quite likely in view of climate change, crop productivity may dwindle further, adding to the woes of rain-fed farmers.

It is well known that the changes in the crop yield depend not only by the changes in rainfall but also by the changes in temperature and carbon dioxide concentration. While positive changes in these two parameters can lead to increase in crop yield, the impact of temperature increase may be negative for the tropics. Doubling of carbon dioxide concentration by 2030 will increase global average temperature by 1-3 degree centigrade an indicating a decrease in cereal production. A shift in wheat production belt can be predicted, indicating that increased temperature from global warming is likely to reduce the profit from wheat cultivation and farmers of lower latitudes to opt for maize and sorghum, better adapted to higher temperatures. In Rajasthan pearl millet output will fall by 10 to 15 per cent due to increase in temperature by 2 degree centigrade. Acute water shortage conditions combined with thermal stress would adversely affect wheat and more severely rice productivity in northwest India even under positive effects of elevated carbon dioxide level in future. North India has already shown an increase in minimum temperature by about one degree centigrade in rabi season.

The Natural Resourse Management division of the Indian Council of Agricultural Research (ICAR) have observed that medium-term climate change predications have projected the likely reduction in crop yields due to climate change at between 4.5 and 9 per cent by 2039. The long run predictions paint a scarier picture with the crop yields anticipated to fall by 25 per cent or more by 2099. This will have a detrimental effect on farmer's income and purchasing power, with obvious down-the-line repercussions.

Though the rainfall records available with the India Meteorological Department (IMD) do not indicate any perceptible trend of change in overall annual monsoon rainfall of the country, noticeable changes have been observed within certain distinct regions.

Climate change is also reflected in the increasingly fluctuating weather cycle with unpredictable cold waves, heat waves, floods and exceptionally heavy single-day downpours.

The most noticeable of such events in recent years included the country-wide drought in 2002, the heat wave in Andhra Pradesh in May 2003, extremely cold winters in 2002 and 2003, and prolonged dry spell in July 2004 and January 2005 in the north, unusual floods in the Rajasthan desert in 2005, drought in the north-east in 2006, abnormal temperature in January and February in 2007, and 23 per cent rainfall deficiency in the 2009 monsoon season. All these events took a heavy toll on crop output.

Indeed, the silver-lining in this dismal scenario in the National Action Plan on Climate Change, launched in 2008, which aims at developing technologies to help rain-fed agriculture adapt to the changing climate patterns. At least four of the eight 'national missions' started under this programme will have direct or indirect bearing on rain-fed farming. These are the missions on sustainable agriculture, water, green India and strategic knowledge. The ICAR-led national agricultural research system is also conducting research on specific projects under the umbrella programme on climate change.

Apart from the use of technological advances to combat climate change, there has to be sound policy framework and strong political will to achieve this objective. State agricultural universities and regional farm research centres too, will have to play a role in developing local situation-specific strategies for adapting the rain-fed farming to emerging climate patterns.

Emerging Extension Models: Experiences

Extension has to shift from merely communicating techniques of know-how to do- how and from technology transfer to technology application mode. The earlier notion of farmer being at the receiving end has to be substituted by farmer first not the last and therefore, farmer participatory extension is to be promoted and the transfer of technology (TOT) has to be farmer-specific and demand-driven.

The technology in order to reach the clientele-farming community from research system needs a well organized and efficient extension mechanism/system of technology transfer from the source of generation to the farming community. This indispensible role of TOT was though realized by the policy makers and the state/ the federal/central government right from the beginning, but all said and done, in Extension, so far, the major emphasis has been on cereal crops which resulted into green revolution in India. Now, to have not only food security, but also nutritional security and to make farming a commercial business enterprise rather than farming, we have to emphasis on diversified farming and demand-driven and market-driven extension.

Though, it has always been said that no one method of extension education can be fully effective and hence a combination for wider impacts in less span of time yet in actual practice, extension personnel ignore this approach and they don't use combination of methods. It is being observed that basic principles and methods of extension education are not followed by the extension machinery. To illustrate, the principles of planning from below *i.e.* bottom-up planning, grassroot organizations *i.e.* farmer participatory approach, whole family inclusion, learning by doing etc. are often neglected and ignored. In restructured extension systems these principles should be rigorously followed for creating motivation, desire for behavioral change in desirable direction.

It is beyond the capacity of any extension system to come out with a unified model to resolve all problems of such a complex and diversified agricultural system prevalent in the country. It is recognized that approaches like farmers participation, institutional linkages, system management approach, policy reforms, capacity building, empowerment of farmers and farm women, use of media and information technology and a host of other suggestive approaches may go a long way in making extension system more vibrant. We can perceive market oriented extension system as a sub-set of overall economic development strategy as: Economic development strategy, Agricultural development strategy, Extension system and Market oriented approach.

The desired goals of Indian agriculture should include self-reliance, food security, nutritional security, conservation of biodiversity, environmental safety, risk minimization, sustainability, export orientation, growth and stability etc. The available resources at our disposal to achieve these goals include human, land, water, energy, agricultural research system, agricultural extension system, agricultural infrastructure, input delivery system, output marketing etc.

The broad agricultural activities may consist of crop production (all commodities), livestock production, agro-processing and agricultural services.

The following steps are suggested for market oriented agricultural extension strategy:

- ✰ Identify all agricultural activities which are compatible to natural and man made resources of the area.
- ✰ Short list agricultural activities which are compatible to natural and man made resources and satisfy as many as possible stipulated goals.
- ✰ Promote such short listed activities with technological and other extension support.

The agricultural development strategy in the present context must ensure adequate interface of the resource base including market infrastructure with the agricultural activities while striving to achieve set desired goals. Such an approach may ensure sustainable development of agriculture wherever market orientation is a feasible option.

What is needed is a holistic combination of verbal, written and visual methods in proper proportion to suit the targeted client groups and their suitability for them based on researches/empirical studies on effectiveness of those methods. This will be of great help in ushering into demand-driven Extension/TOT.

ICT Enabled Models of Technology Dissemination

The need for improved agricultural extension throughout the developing world has never been greater. Agricultural and rural development and hence rural extension continue to be in a phase of transition in this part of the world. The vulnerability of farming in the developing world to climate change, to changes in natural resources quality and lack of coping and adaptation strategies at micro and macro levels of decision making are all well documented, while globalization of commodity trade offers a mix of opportunities as well as challenges. The role of extension and support systems in this background is undergoing profound changes while no unified alternative framework has emerged.

Information and Communication Technology (ICT) is an umbrella term that includes computer hardware and software, digital broadcast and telecommunications technologies as well as digital information repositories online or offline, and includes contemporary social networking aspects, read/write interfaces on the web besides file sharing systems online. It represents a broad and continually evolving range of elements that further includes the television (TV), radio, mobile phones and the policies and laws that govern the widespread use of these media and devices. The term is often used here in its plural sense (ICTs) to mean a range of technologies instead of a single technology.

From the perspective of agricultural knowledge and information systems (AKIs), ICTs can be seen as useful in improving linkages between the research and the extension sub systems. The experience of rural telecenters in the developing world shows that ICTs can help in enabling rural development workers to gather, store, retrieve, adapt, localize and disseminate a broad range of information needed by rural families. The ICTs in extension can lead to the emergence of knowledge workers that will result in the realization of a bottom-up, demand-driven paradigm for technology generation, assessment, refinement and transfer.

Extension agent to farmer ratio in India is estimated at 1:2,000. Public agricultural extension services were criticized for being technically weak, providing insufficient coverage of and contacts with farmers. Less than one-third of the technologies generated by Agricultural Universities and ICAR institutes in India were transferred to the farmer's field due to the lack of an appropriate extension model. Direct contact by agricultural experts with all the needed farmer clients cannot be established practically with the available technical manpower and budget in India. Hence the research challenge is to identify an effective means to provide quality and timely technical advice to all the needed farmers using the available experts and their time efficiently. Developments in ICTs offer ample opportunities to accomplish this challenge. Given the complex nature of agriculture and the challenges being faced, the use of multidisciplinary expertise is more appropriate to address agricultural information needs to empower farming community. Rationale is that technology transfer efforts in agriculture sector must harness the huge potentials of ICT to provide better linkage between agriculture experts and farmers for timely and appropriate technical advice to enhance agricultural productivity and improve living standards of farmers in the region. The linkages between agriculture and climate are pronounced and often complex. Crops and livestock are sensitive to change in both positive and negative ways. Agricultural systems are most sensitive to extreme climatic events such as droughts, floods and hailstorms, and to seasonal variability and changing rainfall patterns. Against this backdrop, farmer adaptations are influenced by many factors, including agricultural policy, prices, technology research and development, and agricultural extension services. The poor often bear a disproportionate burden of direct damage from catastrophes and climate change as concluded by most studies in developing countries.

The role of inadequate institutional support is frequently cited in the literature as a hindrance to adaptation. Adger and Kelly (1999) and Huq *et al.* (1999) showed how institutional constraints and deficiencies affected managerial capacities to cope with anticipated natural events.

Extension pluralism is at the core of farmer adaptation strategies and ICTs can offer new advantages in enabling reliable and rapid access to expert information support which is much needed in the realization of adaptation strategies on a large scale.

Leaders in extension research have pointed out the importance of learning from the deployment of contemporary ICTs in extension. A number of pilot projects in applying ICT in rural development are in progress in many parts of the world. Only a small number of these have a bearing on agricultural information sharing and extension.

Gender Perspectives in Agri-Rural Development

Rural women in developing countries are responsible for production of more than 55 per cent of the food grains and comprise 67 per cent of agricultural labour force and 68 per cent of the developing countries population. According to 2001 Census of India, there are 495 million women (48.27 per cent of the total population) in India. Out of the total workers population, female workers comprise 22.5 per cent of which 68.89 per cent are marginal workers. About one third population of women (out of total population of women) are actively engaged in agricultural activities and play important role in agricultural production. It is a well known fact that women's work remains largely invisible and under recorded. The real issue, therefore, is more serious *i.e.* despite their involvement in agricultural work in such a huge magnitude, they have not been actively involved in main stream of agricultural development and there is hardly any appreciation and recognition of their extensive contribution. By and large, they have remained as "invisible workers".

Despite their role as a backbone of food production and provision for family consumption in developing countries, women remain limited in their access to critical resources and services due to cultural, traditional and sociological factors. They need to be given access to credit, land, agricultural inputs, education, training and extension services, decision making and research and appropriate technologies.

Recognizing women's role as farmers and producers of crops and livestock, as users of technology, as active agents in marketing, processing and storage of food and as agricultural labourers, the National Agricultural Policy (NAP) has highlighted incorporation of gender issues in agricultural development agenda.

The major reforms in Agricultural extension at the state level have also been proposed during the 10th Five year plan with 'Mainstreaming Gender Concerns in Agriculture' as one of the suggested reforms. Under the proposed reforms, being executed in 252 districts of the country, the States are being encouraged to prepare State Extension Work Plans (SEWP) encompassing all extension activities they propose to undertake. These Work Plans essentially include number of activities directly benefiting women farmers to an extent of 30 per cent of total programmatic resources and benefits. The efforts are further being made to introduce the concept of 'Gender Budgeting' in Agriculture Sector for which a 'Gender Budgeting Cell' has been set up in the DAC. The core activity of the Cell will be able to add gender dimension to public programmes and policies and bring out a separate chapter on 'Gender Perspective' in Performance Budget and Annual Report of the Department.

Though women are involved in almost all agricultural operations, yet they have inadequate technical competency due to their limited exposure to outside world. This has compelled them to follow age old practices with low efficiency. The strategic interventions for empowering farm women can include:

- ☆ Creating awareness, knowledge and confidence in farm women
- ☆ Rapport building with farm women by econo-technocrats
- ☆ Programme to eradicate gender biasness among rural societies
- ☆ Access to appropriate technology and financial resources
- ☆ Ownership to productive assets
- ☆ Reducing farm drudgery through income generating projects
- ☆ Generating women's self help groups and cooperatives for their own networking and empowerment through group formation
- ☆ Enhancing organizational and social leadership skill of farm women in community action
- ☆ Building their capacity through collectivization and socio-political participation
- ☆ Providing courses through open and distance education mode.

Agricultural Policies and Support System

Many observers of rural development in recent times have commented on the frequent manifestations of unsatisfactory extension performance. Feder *et al.* (2001) have suggested interrelated characteristics of extension systems in the developing world that jointly result in deficient performance, namely low staff morale, reduced efficiency and financial stress etc. One more such key factor is the number of clients and the vast spectrum of information/services needed to be covered by extension systems. Policy makers in the developing world have reacted to this with the deployment of more extension personnel which has continued the emphasis on a more centralized, hierarchical and top-down management systems. The requirement for combining a bottom up approach with the conventional extension process is yet to be fulfilled and the limitations on the extension process to influence issues such as credit availability, input supplies, market linkages and logistics facilitation continue without change. In effect, there has been no visible impact due to such changes within the extension system in many parts of the developing world.

The NARS – the world over have concentrated on agro-climatic conditions and agricultural characterization for their R&D scheme of things. They have come to realize late that, many scientists are still in the process that, together with agro-climatic conditions, the socio-economic circumstances of the farming communities are equally important for need-based and appropriate technologies. The poor and slow spread and poor adoption and diffusion of technologies are partially because of this mismatch. The rainfed agriculture has made it more obvious, and in fact arduous.

If agricultural research has to become more and more oriented to meet the continuing growth of food needs in developing countries, it should consider strong role of social sciences in priority setting, technology assessment, recognizing role of users in dissemination and adoption, monitoring and evaluation, partnering and policy formulation. More importantly, agricultural research should not only be limited to expanding food production per capita, but should also be able to enhance ability of individual to secure and be guaranteed food security.

Most research institutions are commodity oriented rather than a systems or livelihood oriented and reductionist, not holisitic. Issues linked with human resource development and lack of well designed and well maintained research facilities do also contribute to under-performance of national agricultural research systems in developing countries. To meet the socio-economic and agricultural challenges of the 21st century, the government of developing countries have to concentrate and address on the following issues:

a) Need to recognize that technological advances alone are not enough to fight hunger
b) Need for comprehensive systems
c) Need for more local people's perspective – Farmer led research and extension
d) Need for more consideration of gender issues
e) Need to address agriculture – health relationship (Nutritional Security)
f) Need for policy, institutional and organization reform and
g) Need for innovative funding sources and public private partnership.

A range of extension initiatives from the public and private sectors that explain the way extension agenda is expanding as embodied in the concept of "extension plus" have pleaded for new experiments in extension. Pluralistic institutional arrangements are emerging and are finding wider acceptance and this is mainly because developing countries have realized the need for extension to engage in a wider range of issues beyond merely disseminating production-oriented technologies.

Chapter 2

Agricultural Extension Challenges for Demand Driven Strategies

C. Prasad

Preamble

Even after more than six decades of Independence of the country, the agricultural Extension Education is at the cross-road. Why? We leaned towards foreign assistance, especially the USA. Perhaps that was necessary: (i) in want of resources as well as (ii) experience in the agricultural extension system/sub-systems for sustainable agricultural and rural development. In 1990s the last extension mechanism of T and V extension system of the World Bank completed its term and we were left with a vacuum in this respect. We did not prepare for it in advance. This is why, ATMA the main extension agency of the Government is still in the infancy stage, being spread and stabilize in all the districts. Thank God, the First-line Extension System of the Indian Council of Agricultural Research (ICAR) being implemented through SAUs and the ICAR Research Institutes, is holding the extension education flag high, but it can hardly replace the State extension system, which is mainly responsible for the national extension programmes and activities. Where we miss the boat? This paper intents to examine the basic issues and discuss them for future posterity.

In all these enquiries, there are three dimensions of the challenge: first, is the structural design – organization, ATMA is the emerging one; second, the functioning of the system, where the commitment of the extension personnel and the subject-matter specialists are at stake; and third, the agricultural Extension Management, which is the missing link – a very weak area to be reckoned with. Our doing part has to be improved immensely in order to see that the system works towords the ultimate goal – raising agricultural productivity for food and nutritional security of the population with social justice – 'UNTODAYA' of Mahatma Gandhi has to be remembered and acted upon for the poorest of the poor. Hunger and poverty has been the bane for the nation; also for the developing countries.

We need to learn from the successful experiences of the developed countries especially the USA where the 'Cooperative Extension Work' has been a successfully sustained model since 1914 with the Smith Lever Act. America, being the nation of the nations, acquired matured and experienced nationals who came from different European cultures happily devoted to agriculture; educated their children who stayed on farms; and the commitment to work and agriculture was total. The German nationals contributed a lot in this endeavour; They were enterprising, adventurous and innovative – developed all facets of agriculture gradually, steadily with their hard work and engenuality.

Conceptual Clarity

There two organic aspects of agricultural extension: (i) knowing agriculture, agricultural technologies, and agricultural extension; and (ii) the application of these knowledge and technologies through extension education strategies and approaches. In order to be successful, the extension professionals including subject-mater specialists must 'Own it as if the extension is their own part and parcel, which is devoted to the great cause of – food and nutrition security for the national; poverty alleviation; and employment generation. The following concepts, the extension system must insist upon for the field programmes and actions.

- ☆ Defining Extension: The basic discipline for TOT is Extension Education which when applied to agriculture is Agricultural Extension, to Veterinary is Veterinary Extension or to Industry is Industrial Extension. Education is the keyword % all TOT efforts, no matter who does it (ICAR/SAUs/State Departments, etc.), are educational in nature/character: education, training, demonstrations, fairs, exhibitions, use of mass-media, information technology – all are ment for education and motivation of the farmers which may lead to acceptance of recommended technologies, and then finally the adoption by the farmers as well as diffusion in the farming communities.
- ☆ Agricultural extension services are inputs support (seeds, fertilizers, chemicals, pricing policy, crop insurance, credit policy, land reforms, state farms management, soil and water testing, etc.), are the normal responsibilities of the State Governments, in addition to the extension education function. In reality, the state functionaries are doing mostly the extension services rather than playing the educational and motivational roles.
- ☆ Resources Distribution: As per the policy in the ICAR/SAU System, 10 per cent of the total budget and 20 per cent time of the scientists are expected to be spent on TOT. This is a reasonable proposition, but is this happening? If we add to this, the extension education role of the Ministry of Agriculture/ State Governments (exclusive of services role), the budget, normally should be more than one-forth of the overall total agricultural budget? Are we

investing that amount for the extension education programmes? There is a big gap. Some governments in its 10th Fifth Year Plan spent only 2.5 per cent of its budget for agricultural extension, could you imagine?

- ☆ Extension Functionaries: There are two categories of staff who are directly involved in the Extension Education Programmes like the two sides of the coin:(i) Extension Education Professionals and workers; and (ii) the Subject-Matter Specialists (SMSs) – Agronomists, Horticulturists, Entomologists, Animal Scientists etc. This dimension of TOT has to be understood by both the professionals % one can not do without the other. At present, the common understanding is that TOT means the job and responsibilities of the extension professionals only or if at all, the SMSs are the light support to extension programmes without owning it as their own programmes or responsibilities. This dichotomy has to be done away with so that both feel responsible to the extension programmes and they together share the credits or otherwise accruing out of these activities. Both, in fact, should form a Cadre of Agricultural Extension in the state as a long-term strategy.
- ☆ Specializations in Extension Education: All disciplines have specializations for developing competencies in specific areas. So far, we have not done much about it in agricultural Extension or if we have done it, they are not visible and effective. We may develop specializations, besides general extension education courses, in (i) Agricultural training including vocational and entrepreneurship development; (ii) Agricultural Extension Systems and management; (iii) Research methodologies in social sciences including a thorough knowledge of parametric statistics; (iv) Agricultural Extension Approaches and methodologies including the ICT–Information communication technologies as well as the mass-media; and (v) Women in Agriculture.

These conceptual ideologies and framework must be clearly understood by our extension agencies, the Extension Professionals as well as the Subject-Matter Specialists (SMSs), and adhered to them in practice in order to be effective in the field programmes – promotion of adoption among the farmers and diffusion in the rural/ farming communities.

Why Skewed Development?

We are talking about socio-economic development in country almost like a slogan and rightly so. But what is happening on the ground? Only economic development–raising productivity, production and increasing GDP (Gross Domestic Product) –a capitalistic approach. The social dimensions of development –alleviation of poverty and hunger, social and cultural cohesiveness, social values etc. have been the missing links. The result is that, we have become self–centered and are crazy for money and accumulation of wealth even at the cost of the poor and weaker. How poverty and hunger could be eliminated from developing countries, if this is the social environment? The millennium goals of the U.N. thus have remained almost where they were. Richer are getting richer creating further gulf between rich and poor. From economic power –we are getting hold of the political power – more unchallenged money and hence a lot of money laundering scams especially by the politicians and the powerful. See the newspapers daily in developing countries–half the papers are occupied by corruptions and scams. Unlimited wealth with people, unless socially conscious and enlightened, corrupts the richer and political power corrupts them absolutely.

The societies are disintegrating, people are getting individualistic, poor, weaker and minorities are ignored and thus the results are emergence of Naxalites, Maoists, and terrorists. One could imagine and visualize how the future of the societies would look like; we are leading towards unprecedented crisis and chaos, ultimately a civil war – the devastating destruction may be. Seeing this extremism and injustice, God is also not kind – He sends tsunami like phenomenon; climate change and global warming have started showing their heads. No doubt, we are leading towards perils and perish, a point of no return. Our Indian mythology (GEETA) says that "one should acquire only an optimal amount of wealth enough to meet their basic needs; and if this is not done, the hungry people will snatch away the wealth for their survival – it is not sin"! This has started happening here and there now.

The root cause of hunger and poverty, in some sections of our society, is due to excessive wants, desires and greed of the rich people. Rich is getting richer and poor

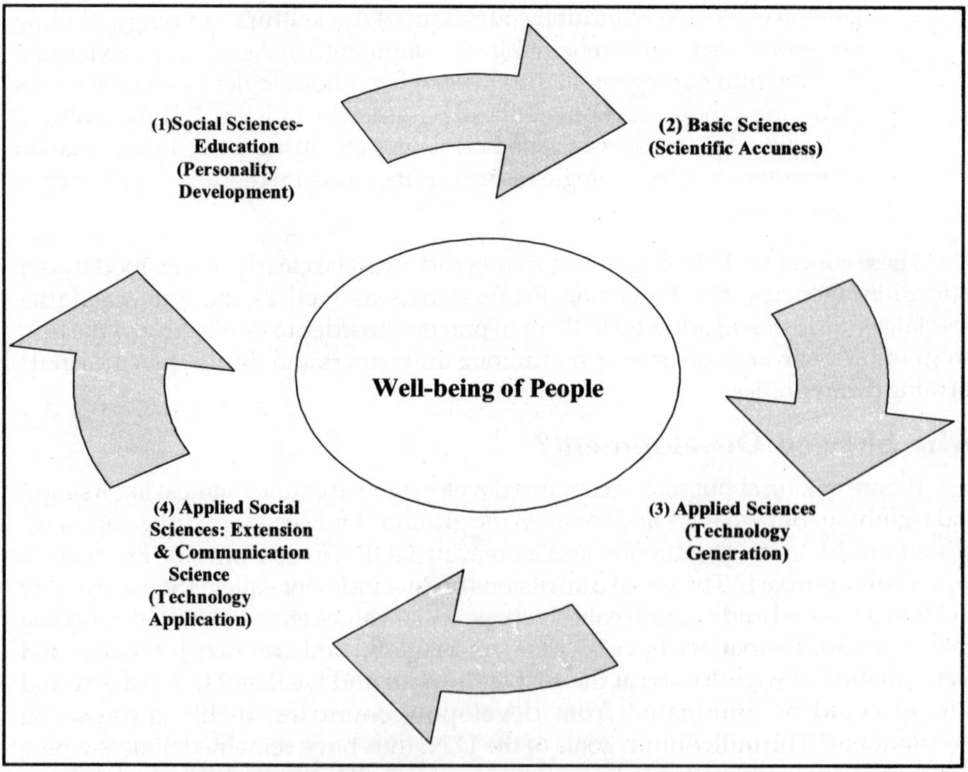

Figure 2.1: Priority Place of Social Sciences Education Over Basic Sciences: The First.

the poorer, is the common slogan since long. This had been happening gradually in the countries like a slow poison in the twentieth century and on the on-set of the twenty first century. The aberrations of the society are getting more visible in terms of class, caste, creed, discrimination, deprivation, poverty, hunger, violence, chaos and calamities. Societies are disintegrating, whereas Baha'i rightly thinks of 'One Soul and One Country! The King of Bhutan, (His Highness Jigme Singye Wangchuk) has launched the concept of Gross National Happiness (GNH) *vis-à-vis* the Gross Domestic Product (GDP). The agricultural economists and allied social scientists in the USA and elsewhere have started studying this concept and approach. This has the angle of general social well being of the people, rather than promoting an unbalanced approach to development in favour of ever multiplying wealth (Figure 2.1). Today, the ratio between rich and poor is 1:74 against 1:30 as it was in 1970s. The good plants are grown on a conducive soil culture. Likewise, diffusion of technologies, may that be agriculture or industry, would take place on the foundation of conducive social order. But we want accelerated promotion of science and technology in the societies which are suffering from many social ailments and maladies – suspicion, distrust, selfishness, and the least respect for the poor and the peace. Therefore, our approach to development should be to prepare the rural or urban societies socially conducive, united, receptive and rich – a good and effective ground for absorption of technologies for sustained development, rather than wishing miracles to happen in an unconducive social order. This is a deeper issue which we have to understand and work together for the agricultural research and development; this is equally relevant to industries and other allied sectors of development and research. Human resources must be socially alive; it is our Job to educate and change them, though it will be a gradual process.

A study, Limit of Growth (1970s), commissioned to Massachusetts Institute of Technology and Boston University, USA, stated: "Social problems apparently created by technology cannot be solved by the application of yet more technology; social problems in terms of social value systems and values have to change fundamentally for the problems to be solved" (ICSSR, 2001). The social core issues recognized by the U.N. system are (i) poverty reduction, (ii) human rights, (iii) peace and security, and (iv) the advancement of women. The eight millennium development goals of the UN were framed around these basic issues.

"Nobel laureate Amartya Sen has so definitively demonstrated that empowerment leads to entitlements and entitlements lead to enrichment. The skewed distribution of the fruits of the impressive progress the country has made requires immediate course correction because economic progress in itself cannot be our national goal."

Why skewed development may be seen in our educational systems, both general or professional? Because social Sciences have been ignored in favour of basic Sciences – the job and package oriented education. Education is a Social Science as per the knowledge classification (Social Sciences) of the Dewey Decimal Classification Scheme (21 Edition), U.S.A, 1996, which is basic for personality development including social values, morality, commitment and character ethics. A drastic change in educational system is due. This must start now, for education has a long gestation

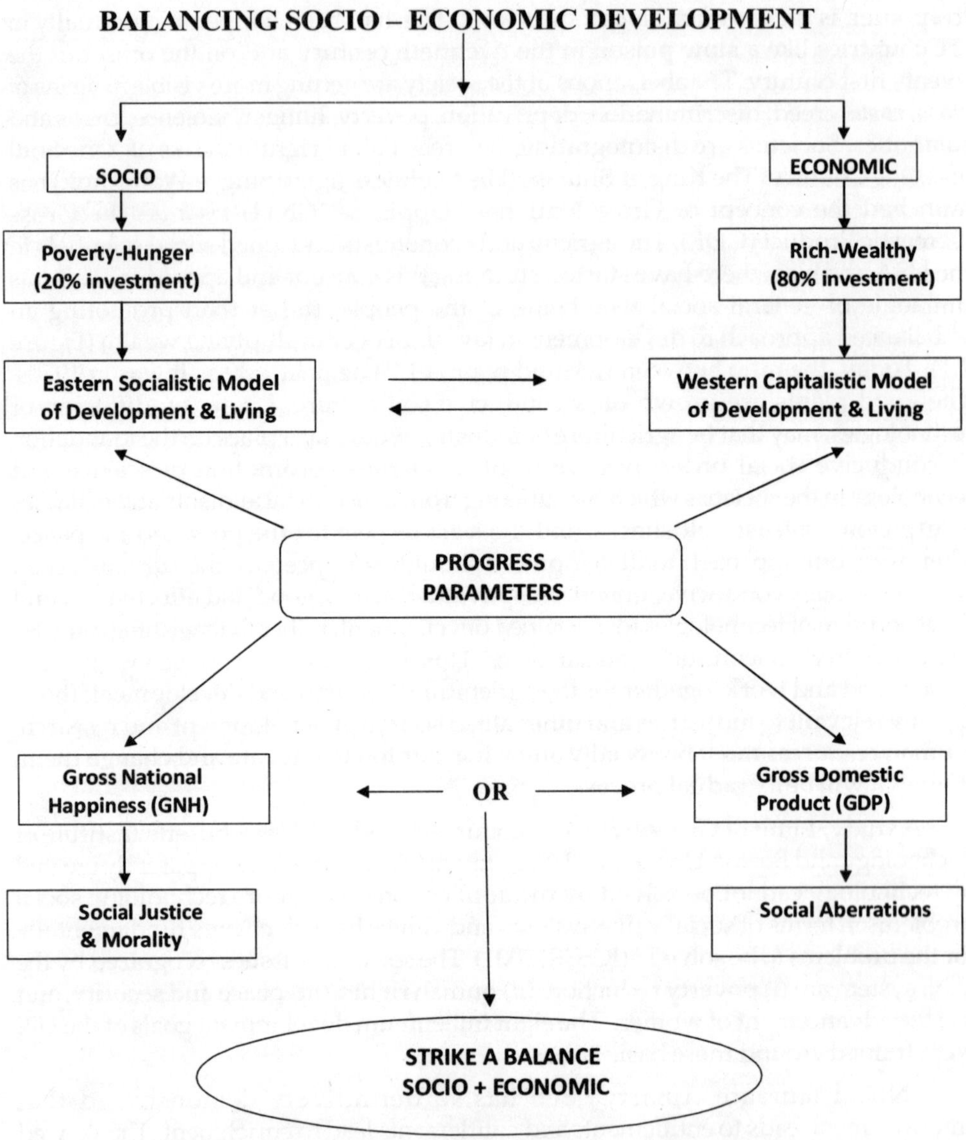

Figure 2.2: Emphasis on more Economy than Social Concerns.

period. A balance education must be promoted in our educational systems; this is in short supply Figure 2.2 explains. What could be the role of extension professional in this respect, we must ponder on this?

Training as the Fourth Pillar

We recite Teaching, Research and Extension – the three pillars almost like a slogan. Now time has come to recognize the 4th Pillar – the Training. Training conceptually is the practical education – skill and job–oriented; one of the devices for TOT. By adding a modest School of Training/Directorate of training in SAUs and deemed universities of the ICAR with only a half-a-dozen core faculty members, and recognizing all the faculty-members of the SAUs and deemed universities as trainers, we can contribute tremendously to the HRD sectors in agriculture – pre-service, in-service, vocational and entrepreneurial – a factory for job creation by offering hundreds of technical courses in all agricultural disciplines including certificate and Diploma Courses. The Mentor Group of Bihar recent Agricultural University, Sabour, bhagalpur headed by Dr. R.B. Singh has recognized this fourth pillar on its merit. This is being implemented in the newly created BAU.

We are excited about having established nearly 600 KVKs, but in view of our mere size of population – agriculture and non-agricultural, we are far behind the required numbers; quality of training is another factor. The National Commission on Agriculture (1975) recommended for at least three KVKs per district in view of the size of the district both areas and population-wise. We have just started the second KVKs in the larger districts only. Let us look at a few international examples in this context:

- ✰ Germany, population 86 million, has 100,000 Vocational Training Institutes for a work-force of 42 million and nearly 500,000 companies are involved in practical training as well as handling of apprentices. About 200,000 professional experts from different sectors of the economy are registered and are used for imparting theory and practical hands-on-training on about 3,000,000 people per year.
- ✰ Even a small country like Austria, population 8.5 million, has 5000 centers, training 300,000 people per year and 38,000 companies provide practical training. No wonder that they have the lowest unemployment rates and one of the highest labor productivity in the EU.
- ✰ In the USA and the EU nearly all Vocational Education and Training is in the PPP (Public-Private-Partnership) mode. About 70 per cent financed by the private sector or employer's organizations/local community and 30 per cent by the Central and State Governments.
- ✰ China has 500,000 vocational institutes, out of which nearly 70 per cent are in rural areas. The total number of people trained within China is about 60,000,000 people per year. It is, therefore, not surprising to note that China with arable land much less than India produce's nearly 100 per cent more foodgrains, although, it has very harsh climate.

Social Education – the Weakest Link

The human resources development in any country is a crucial and critical issue. A good education means developing a balanced personality combining social parameters as well as the economic dimensions. Today this lopsided in most of the Asian countries including India; it is more 'economic' than 'social.' Hence, perhaps, inadvertently we are preparing people for money and wealth as a primary goal, devoid of social justice. It is money, money and more money and power, more power and absolute power. It seems there is no end to desires and wants, whereas Mahatma Gandhi (India) advised us to have control over excess wants and unnecessary desires.

Education is one of the social sciences according to Dewey Decimal Classification of the USA (2001); social sciences deal with 'man' and 'society'. Education is basic to 'personality development' including 'social' and 'economic' traits. To make substantial contributions to society, these traits must be balanced and judiciously used in the interest of the society at large. But 'skewed development' has been the order of the day globally – more economic than social, leaving one- fourth of the population in developing counties under hunger and poverty. The West has followed the path of 'capitalistic' pattern of society mostly devoid of thrust on social values and the East is crazy copying the West ignoring the 'socialistic' pattern of society. Thus rich are getting richer creating gulf and divide between rich and poor; latter in share frustration are being converted into Traitors, Naxalites, Maoists, Decoits, and Kidnapers – a danger to the Society. In Indian mythology, *Bhagwat Gita* stated: 'One is entitled to only that amount of wealth which is enough for ones upkeep; the rest has to be shared and if one does not do so the same can be taken away'. Thus, are we not leading towards a volcanic social eruptions between the two divides? While several approaches and methods could be adopted to contain this baffling problem, the educational system must change for the better with balanced doze of social sciences vis-à-vis basic sciences producing socially concerned and technically committed human beings (Figure 2.3).

Education has a long gestation period, almost 20 years upto highest education of these years in the first half, social sciences should be amply focused together with the art or science education; the other half becomes the professional courses, more and more specialized. But initial social sciences input in socially constructive education will last for ever; a balanced education for balanced education for balanced personality development will remain balanced all through producing a good human beings, good citizens the ultimate goal of life and living (Figure 2.3).

Extension education is out of school education, informal or non-formal education – a more difficult education indeed. This is why, we have to put in a lot of efforts to realize this educational needs and prepare farmers as enlightened, progressive and innovative farmers. The farmers should be encouraged to educate their children and retain them on far as best as possible. Farming today are more paying and more dignified than menial jobs in cities.

Demand-Drive Extension Education

ATMA as Field Extension Design

ATMA (Agricultural Technology Management Agency) was a product of the NATP, which was designed to the Integrate all extension agencies at the district level in order to strengthen the field extension work. Nevertheless, in absence of T and V extension system, it was lately assigned this role of Field Extension Education/ Services by ATMA through the mechanism of Farm Schools on the farm of the progressive farmers recommended by the National Commission on farmers (2005).The concept of running 'Farm Schools' through ATMA was an innovative mechanism,

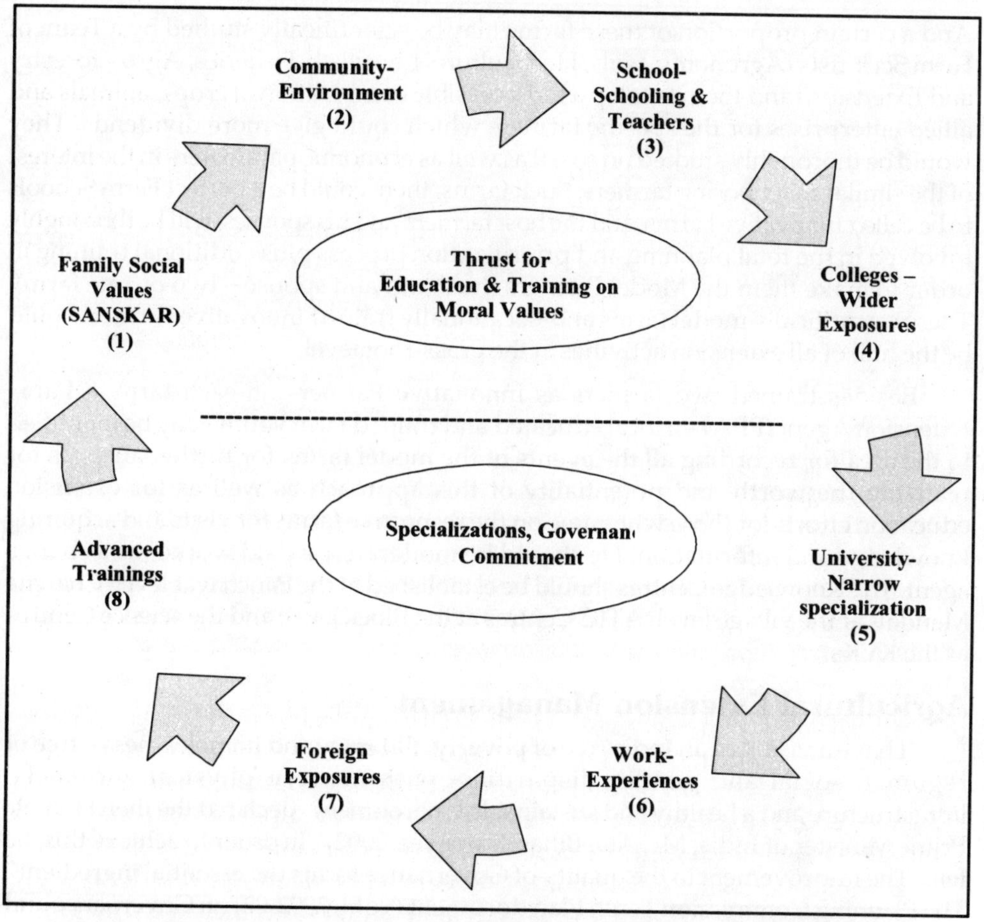

Figure 2.3: Informal and Formal Education for Morality and Commitment.

no doubt, but it suffers on two major accounts: (i) such farms have not been developed on a Farming System Mode (FSM) and (ii) the farming is not based on a study and analysis of the natural resources of the Farms. – Perhaps they have been taken on the face value because they are relatively better than the others. Whether they are effectively functioning or not is another story. The Farming System Mode will be that farming which is based on the potentiality of the natural resources (soils, water, rainfall, climate etc), the right combinations of crops, horticulture, livestock, fisheries and agro-forestry recommended and practiced, which could be always better productive than only one crop or two in the kharif and Rabi seasons.

Model Farms and Innovative Farmers

The ideal and the most potential proposition will be to identify certain potential farms (marginal, small, semi-medium, medium, high) strategically centrally located, on the proportionate basis according to the numbers of operational farm holdings.

And a certain proportion of these farms may be scientifically studied by a Team of Farm Scientists (Agronomy, Soils, Horticulture, Livestock, Fisheries, Agro – forestry and Extension) and then develop a most feasible combination of crops, animals and allied enterprises for the specific farmers which could give more dividends. They would be thoroughly studied on social as well as economic parameters in the interest of the similar neighboring farmers. Such farms, then, could be a perfect Farm Schools to be called Innovative Farms, and the host farmers and his spouses, will be thoroughly involved in the total planning and prioritization process plus additional training in order to make them the Model Farmers (husband and spouse – two on one farm). These scientifically model farms and vocationally trained Innovative Farmers could be the hub of all extension activities at the grass-root level.

Besides trained two farmers as Innovative Farmers on each farm, a Para-Extension Agent (PEA) – a local educated and trained farm youth, may be identified in the area for recording all the events of the model farms for further analysis for realizing the worth–and potentiality of this approach as well as for extension education efforts for those who came on the respective farms for visits and acquiring knowledge and information. He should be moderately a paid worker as a change agent. The Knowledge Centres should be established at the Panchayat level, Charcha Mandals at the village level; ATIC Centres at the Block level; and the science Centres as the KVKs.

Agricultural Extension Management

'I have a vision of an India free of poverty, illiteracy and homelessness – free of regional, social and gender disparities- with modern physical and social infrastructure and a healthy and sustainable environment" declared the then Hon'ble Prime Minister of India, Mr. Atal Bihari Vajpayee(2002)'. In order to achieve this, he felt, 'The improvement in the quality of Governance forms the essential ingredient'. The Planning Commission Tenth Plan document (vol.I, 2002-07) on Governance and Implementation asserts: "The issue of governance has in the recent times emerged at the fore front of the development agenda. Good governance is one of the most crucial factors required if the targets of the Tenth Plan are to be achieved. It is also this factor, or rather lack of it, which could be the cause of immense disappointment and missed development opportunities'.

Management is a universal and omnipresent discipline – all of us are using it every where, consciously or unconsciously, either individually or collectively. Unwittingly though, management has been treated mostly by the professionals only in the context of an organization or institution.

Management is a new frontier in agriculture. Agricultural extension management is one important aspect of it. Management is a social process of working with people or tackling people in an organization most efficiently, economically and sustainably. It creates a conducive working environment in an institution for greater participation, team work and production through synergistic effects. Agricultural development systems represent a complex organization including research, education, extension and training functions and varied clientele groups. The multiplicity of institutions and agencies serving agricultural sector and sub-sectors with heavy investments demand an efficient management system.

Figure 2.4 gives a broad dimension of Agricultural Extension Management System. There are five functional areas – research, education, extension and training in extension discipline and two clientele systems where extension management has to contribute. For instance, in research in extension system we are concerned about (i) generation of appropriate technologies (AT) by the subject-matter research scientists in association with extension system and (ii) research in Transfer of Technology (TOT) by the extension and social science professionals; in Education for providing qualified and quality extension specialists (ES) as well as well trained subject-matter specialists (SMS); in extension working with two extension systems : (a) First-line Extension System (FLES) and (b) the General Extension System (GES); in training with two important areas like professional training (PT) (In-service) and vocational training (VT); and the clientele system should include not only rural areas, but also urban extension where agro-industries and corporate sectors in agricultural development are important. In addition, technology transfer in developing countries now take an important place in the scheme of globalization of agriculture, and the creation of World Trade organization (WTO) with its world wide mandates.

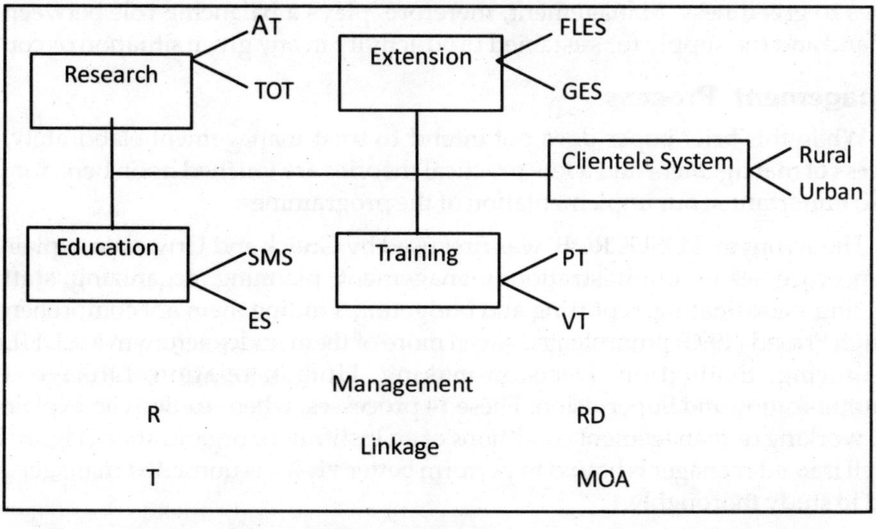

Figure 2.4: Dimensions of Agricultural Extension Management System.

A student does the time budgeting in his study, sports, and leisure-time activities. A saint manages to live in a detached manner from the worldly allurements and keep on organizing himself to serve the cause of God. A husband manages his family, the football Captain his team, the Principal his college, the Vice-chancellor his university, and an Executive his corporate body. Whether one is a manager or the one is being managed in both the cases an understanding of and insight into the principles and practices of management and administration would help. An incumbent of an institution with such background and experience would know how to contribute to and work with a team of employees in the total management milieu and maintain his working relationship with the superiors.

Today, the promotion or use of partnership and participatory approach has become rightly a slogan of our time. But it has been a rare commodity in most situations, because of the absence of management interest, education and training among the people. Therefore, the time has now come to promote management education as a continuum (management continuum concept or say an universal theory), like any other educational area and not in the organizational or institutional context alone as has been normally the case. We must build a management culture for the optimization of resources both men and material, in all situations every where. Hulse (1977) refers to the concept of management even in the one-to-one employer–employee relation, though he states, "there was no great need for an elaborate management structure' then (early 18th Century).

The complexity and intensity of management in a given situation always vary in proportion to one's needs and resources, or say, in economic terms, demand and supply. Need is contagious; it is created and has a natural tendency of being gradually enlarged and multiplied. Need is too dynamic a concept which, depending upon the circumstances, either promotes professional excellence or breeds contempt when it comes to greediness. Management, therefore, plays a balancing role between the demand and the supply for sustained productivity in any given situation or context

Management Process

While this brief paper does not intend to treat management elaborately, the process of management and a few practical theories are touched upon here, for they are so important in our implementation of the programmes.

The acronym 'POSDCROB' was first used by Gulick and Urwick to explain the seven processes of administration/management: planning, organizing, staffing, directing, coordinating, reporting and budgeting. Finding them not comprehensive enough Prasad (1993) promulgated seven more of them under acronym MEDHLiCS: Monitoring, Evaluation, Decision-making, Human relation, Linkage, (Li), Communication and Supervision. These 14 processes, when studied can explain the total working or management conditions of an institution/organization (Figure 2.5). A well trained manager is bound to perform better vis-à-vis untrained managers. We need to study thoroughly.

Fusion Theory of Management

This is a practical theory, which the mangers must understand and act upon. It is relatively easy conceptually, but is very difficult in application. But all the same, there is no way the good managers can do without it. This theory is commonly found in operation in private organizations. This theory has a basic assumption, *viz.*, the developmental processes and goals of the individual and his organization are, at crucial points, fundamentally different and antagonistic (Argyris, 1953, Bakke, 1952). In other words, the normal tendency of the incumbents is not to work as much as desired by the organization, but get the best out the organization; and the organization also wants to get the best work done by the incumbents with least investment. This is

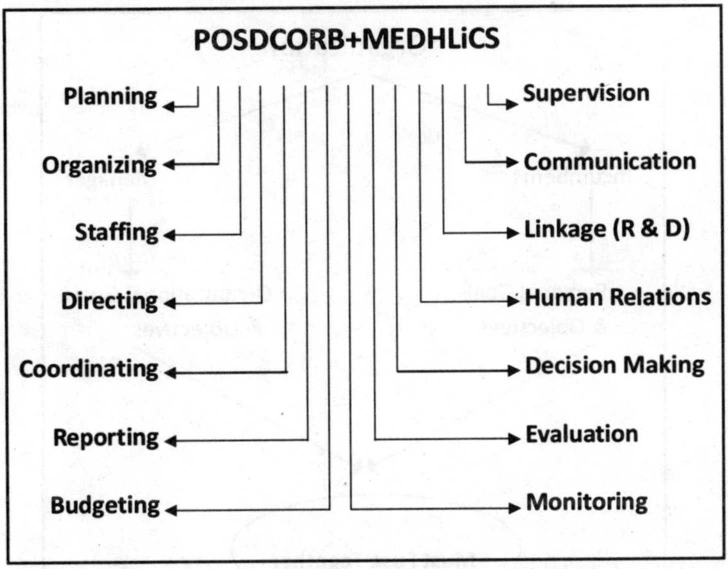

Figure 2.5: Fourteen Processes of Management.

an antagonistic reality. The fusion theory postulates that the incumbents have their own goals and objectives which they want to get fulfilled (salary, promotion, status, and so on), whereas the organization has its own goals and objectives (higher productivity, production and profit), which also must be fulfilled. Therefore, the fusion of the two goals (of the incumbents and the organization) is necessary. The fusion process is said to have occurred when the individual obtains the maximum expression of his personality that is possible and, simultaneously, the organization has its demand fulfilled at the highest possible level. As mangers, we are faced almost a very day with such situations and interaction, and we are resolving them also somehow, but perhaps without the knowledge of this theory. Hence, many times, we are not able to decide where both the interests can converge. The unwillingness to work and the common agitations and strikes are the manifestations of decision-making where both the points of views are not deliberately considered (Figure 2.6).

Need for an Autonomous Council

In spite of over five decades of Agricultural Extension System in the country, we are still not sure of the extension education viability and sustainability of the extension system. The main reason was that we enjoyed the foreign projects, support and experiences and in that process we did not prepare ourselves for our model extension to be with the result today we are on the cross-road. There are a lot of reports, suggestions and guidance by the high level Communities and Commission –NCA, NCR etc. but nothing seems to be working. The reason is obvious; we have lost the track of being sincere and committed to our tasks and responsibilities. Agricultural development demands committed field–workers to be helped and guided duly committed by leaders both professional as well as political. But how are these working, one can obviously see which is not encouraging.

Figure 2.6: Fusion Theory of Management.

Then what could be done under these circumstances? We need very committed and aggressive leadership and management set-ups. At present we have two subordinate offices for extension–one in the Union Ministry of Agriculture, the Directorate of Extension headed by a Joint Secretary (IAS) under the additional Secretary/Secretary and in the ICAR, the Division of Extension is headed by the Dy. Director General, under the Secretary, (DARE) and D.G. ICAR. We need an autonomous Council for Technology Transfer in Agriculture headed by a D.G. in the rank of Secretary, Agriculture.

We have a lot of work on research and technologies; they are not the limiting factors today. Our farm scientists have done exceedingly well , but on the TOT side we are extremely weak. Of the two arms of extension education in the country, the ICAR has done very well, but there role and scope, as per design and policy, are limited . The ICAR is devoted to only for the First line Extension efforts, that is, demonstrations and advisory services on the latest agricultural technologies and methodlogiest. The major responsibilities of the extension work rest with the State Extension Systems where we have the basic problems. The State Extension System, to a great extent, are functioning as the service providers and not as the Extension

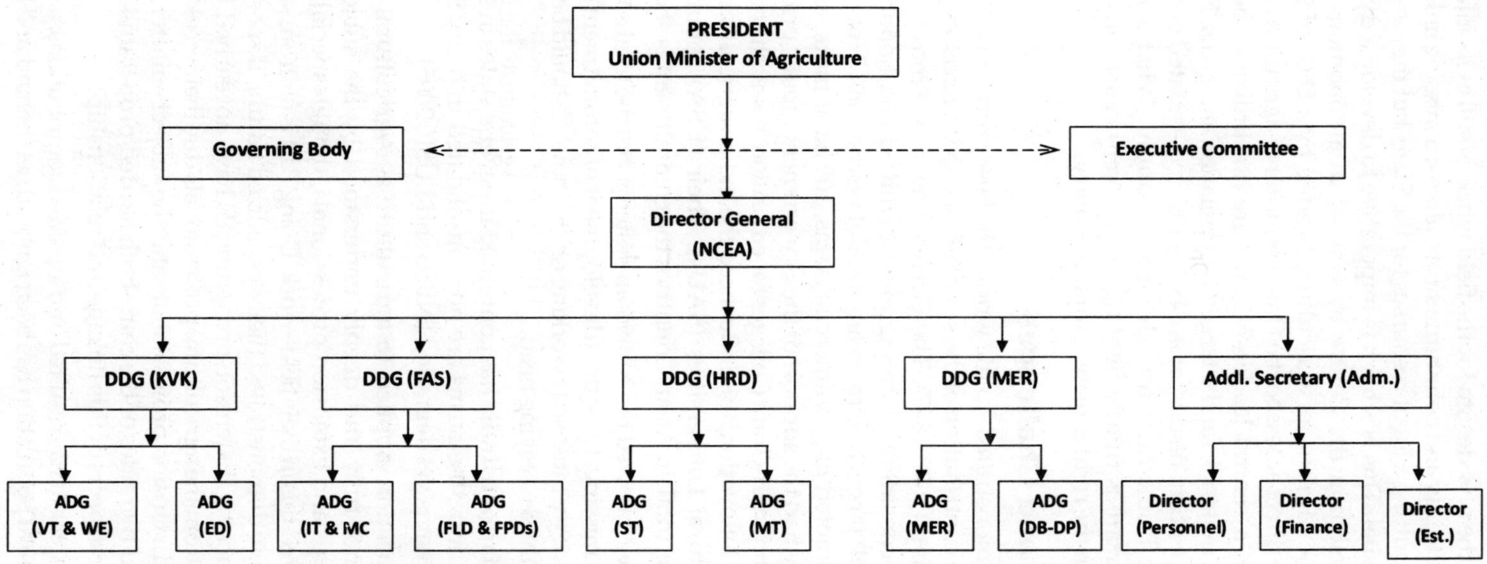

Figure 2.7: Autonomous Council for Agricultural Technology Transfer-Illustative.

FAS: Farm Advisory Services; MER: Monitoring, Evaluation and Research; VT and WE: Vocational Training and Women Empowerment; ED: Entrepreneurship Development; IT and MC: Information Technology and Mass-Communication; FLD and FPD: Front-line Demonstration and Farmers Participatory Demonstrations; ST: Staff Training/In-service Training/Advanced Training Centres; MT: Management Training; DB and DP: Data-base, Documentation and Publication; Est: Establishments.

Educators. The ATMA, the new design for the field work, is still at its infancy stage. And added to that, the sincerity and commitment are almost a missing link. While the Government of India is putting a lot of resourses for the State, but they are not going for the right causes as usual. This is why, it is imperative to develop a dynamic and strong extension leadership at the Centre by way of an autonomous Extension Education Council in the Ministry of agriculture headed by a D.G. of the rank of Secretary, who will have direct access to the Union minister of agriculture and higher ups. In absence of such a central leadership, we are not utilizing the available resources, both men and material, for the targeted population and goals. For the field work ,perhaps, only 20 per cent resources are being utilized as stated earlier by our beloved Hon'rable late Prime-minister, Shri Rajiv Gandhi. What a frustrating situation? Mr. Anna Hazare is rightly fighting for this very cause; no doubt, it is becoming a movement in the right direction for the country.

Areas of Concern and Challenges

1. The U.S.A. demonstrated, to the world, the best productive agriculture, which was the contribution of the so called European refusees – educated and enterprising (1620-1647). The Germans were the most progressive farmers, who considered farming most dignified and honorable. They settled on small farms free from any feudal restrictions. However, since 1975, they promoted big commercial farms, 100 ha or more, an evil start they said. They had the support of the Government inputs, prices and the promotion of the Land-Grant Colleges for education, research and extension –an integrated innovation; this influenced the Indian agriculture by way of State Agricultural Universities (SAUs). Their (USA) commitment to agriculture was total; in Asian Countries the commitment is short supply!
2. In India we have followed the American systems, but the total commitment like the USA is missing; the latter (development of commitment) has a long gestation period, we must start working on this now. It should be one of our priority areas this is a crying need.
3. Feudalism on farms in India has continued in several states in spite of the Ceiling restrictions; the farms are not consolidated in many States; and 146.82 mha remains yet degraded (NBSS and LUSP, 2005).
4. Extension Education, applied to agriculture as Agricultural Extension Education is not fully and deeply understood by the subject–matter specialist groups and extension professionals, though specialized, have not demonstrated this in their field–work. Doing Part is far from satisfactory; they should have demonstrated their specialized nature, if specializations had not been lacking. Extension professionals have not earned the respect they deserved from the agricultural professionals due their weak field work and activities; Extension professionals and the subject –matter specialists must be treated two sides of the coin –both needed proper understanding and the deeper aspects of this linkage and relationship.
5. Skewed Development has resulted into lopsided contributions; a substantial section of the rural population has been ignored, left behind and backward.

Generally people have grown as self–centered, individualistic and societies are disintegrating; capitalistic pattern of development has taken over the socialistic pattern of growth. Our education has been weak –individuals are running for money, wealth and packages. The concept of Bhutan King, the happiness of the nation vis-à-vis the GDP has deeper implications; the earth as ONE Country approach and mankind its citizens of Baha'i (2009) has a far reaching implications for the nations.

6. The recommendation of using the farms of progressive framers as farm school for demonstrations and training has far reaching implications no doubt if they are implementing well. Nevertheless, the Model Farms and the Innovative Farmers concept (Prasad, C. 2009) will be an improvement over the concept of NCF as explained in the text.
7. The KVK (Farm Science Centre) is an innovation, the root being the National Commission on Education (1964) and the shoot being, the ICAR. The total concept of the KVK has yet to be put into practice due to so many reasons. We must insist on those approaches for skill-oriented quality vocational training and entrepreneurship development, and demonstrations of the latest technologies. This grass-root level institution, the KVK could as well be adopted by the developing nations. In order to hasten the process of KVK faster spread among the districts for taking full advantage of this innovative approach, the Union Ministry of Agriculture, should seek the support of the UNDP/FAO or the World Bank to establish them as soon as possible. This will be a real good project, the World Bank should be interested for its larger implications for the developing countries.
8. Agricultural Management is the weakest link in the agricultural development process, agricultural extension management, in India is no exception; perhaps, developing nations are suffering from some maladies. The Planning Commission, Govt. of India is rightly very vocal about it. We should also learn from the Industrial Management of the country (which is far better managed, efficient and effective). To augment this sub-sector of management in agricultural extension, which is a crying need, the Union Government should establish an independent autonomous Council of Agricultural Technology Transfer (CATT) to be headed by a Secretary rank management professional or a technocrat with wider and deeper field extension experiences and management accuness. The re-organization on this line should follow at the state levels then. A High Level Committee of Extension Scientists and Subject–Matter Specialists may go into this policy issue on an emergency basis. The empire building habit of authorities must be sacrificed in the national interest.
9. Extension discipline is made out of several disciplines. Therefore, the extension professionals unfortunately are not very deep in each case of the contributing disciplines–Education, Sociology/Rural Sociology, Educational Psychology, Farm Technology, adult on the base of agricultural education and the agricultural Communication Technology (ICT). Realizing this weakness, we should promote continuing learning–Adult Education

on those disciplines for improving our commitment and efficiency. We must realize this basic truth.

10. The extension professionals and the subject–matter specialists, at present somehow, are not tuned together like the two sides of the same coin. This area needs improvement in the interest of the field extension work and activities; in its absence we are not cutting much ice in our field work; this is extremely important.
11. Management is an omnipresent concept and in that sense a basic critical requirement everywhere; Maximum attention to this is the crying need of the day.
12. As a long-term extension strategy, the subject–matter specialists and the extension professionals should be made into a cadre with the Time Scale so that their valuable experiences are not lost. This is crucial indeed.

References

1. Prasad, C and Babu Suresh, Edited Book; Social Science Perspectives in Agriculture – A Thrust for Integration, VARDAN, New Delhi and IFPRI, Washington D.C., USA, 2008.
2. Prasad, C., Training for Agricultural Development – a Basic functional Area, National Workshop on Manpower Development in Extension Education for Meeting Challenges of Agriculture, NAARM, Hyderabad, India, 1995.
3. Joshi, Murli Manohar, Sustainable Consumption: issues of A Paradigm shift, ICSSR occasional Monographs Series, Indian council of social Sciences Research, New Delhi, India, 2002.
4. News letter of the B a h'a'I International Community, one country, IRAN, 2008.
5. Khanna, Krishan, Chairman – C II skills Task Force, Employment Ist, Paradox – 300 Million, New Delhi, India, 2007-2008.
6. Anonymous, Tenth Five Year Plan, Vol. 1 (2002-2007), dimensions and strategies Planning Commission, govt. of India, New Delhi, 2002.
7. Prasad, C. Farm Science Centres (KVKs), a chapter in, social science perspectives in agriculture, a thrust for integration, VARDAN, IFPRI, USA and ISEE, New Delhi, 2008.
8. Randhwa, M.S., A History of Indian Agriculture in India, Indian Council of Agricultural Research (ICAR), New Delhi, India, 1980-86
9. Prasad, C. Reaching Resource Poor Farm Families with Relevant Technology and Methodology, Paper Presented in a National Seminar, NAARM, Hyderabad, India -1994.
10. Anonymous, National Bureau of Soil Survey and Land Use Planning, ICAR, Nagpur, India, 2005.

11. Anonymous, One Country, Bah'a'i International Community, July-November, 2009.
12. Anonymous, India 2011, Research, Reference and Training Division, Ministry of Information and Broadcasting, Govt. of India.
13. Anonymous, Agricultural Statistics at a glance, Directorate of Economics and Statistics, Ministry of Agriculture, Govt. of India, 2005.

Chapter 3

Open and Distance Learning System in Extension Education

B.S. Hansra and P.K. Jain

Conventionally, land, labour, capital and management are four factors of production. In the competitive globalized environment, not only the capital and infrastructure but also the information, research and innovation, education and learning systems are vital, therefore, knowledge is a core of development process in the present era. Extension education is system to impart learning to rural and form community to improve their livelihood by changing their behaviour (change in knowledge, attitude and skills). Extension education provides knowledge and skills for better management of farming and increases income.

Open and distance learning (ODL) is relatively new in the field of education. It gains prominence only in the last 25 years. This term is comprised of two terms *i.e.* Open Learning and Distance education. Open learning is a philosophy and Distance learning is a methodology in education. Openness and learner-centric is a basic philosophy of ODL. Flexibility and openness in entry and admission requirements, programme structure, and flexible learner support (when, where and how to study), etc. are important reasons for popularity of ODL. Distance education or distance learning, is a field of education that focuses on teaching methods and technology with the aim of delivering teaching to students who are not physically present in a traditional educational setting such as a classroom. It has been described as "a process to create and provide access to learning when the source of information and the learners are separated by time and distance, or both" (Wikipedia). Physical on-site presence of teachers and learners is also recommended in the new generation distance education particularly in technology oriented courses.

There are a number of similarities between extension education and distance education:

- ✰ Extension education deals with study of farmers problems and needs of the farmers. The subject and areas deal with the distance education is also based on the need of the society.
- ✰ Extension education has no fixed curriculum or course of study and curriculum are developed with the help of farmers based on their requirements. Similarly, in case of distance education, curriculum is developed based on the requirements of the learner and choice is given for selection of curriculum.
- ✰ Decision power is given to farmers / or learners in extension and distance education in terms of when, where, what and how to learn.
- ✰ Participation of learners is voluntary in both extension education and distance education (self pacing).
- ✰ In extension education local leaders are used for teaching and In distance learning, the locally available teacher-counselors are used for teaching.
- ✰ Learners are large and heterogeneous in both extension and distance education.
- ✰ Both are flexible in terms when, where, what and how to study.
- ✰ Extension education is more practical and intended for immediate application in the solution of problems. The programmes / courses offered by the distance education are also need based and have immediate application for improving the livelihood.

The open and distance learning (ODL) system is successful in providing the education to the unreached. It is being employed in agriculture as well for suitable curriculum development and delivery. This paper covers the development and delivery of agricultural programmes through cost-effective and efficient ODL to educate farmers, upgrade their technical, entrepreneurial, and analytical skills.

Development of Education and Extension System

In pre-independence India, five agricultural colleges at Pune, Coimbatore, Sabour, Kanpur, Nagpur and Layallpur (Now in Pakistan) were opened in order to impart agriculture education and bring improvement in production technology. Setting of Imperial Agricultural Research Institute, Pusa (Bihar) and Imperial Council of Agricultural Research were the major initiatives for agricultural research, education and extension activities in the country. The education system has undergone lot of changes in post-independent India. The Imperial word was replaced by Indian Council of Agricultural Research (ICAR). State Agricultural Universities (SAU) were established on the philosophy of Land - Grant College of USA. Presently, there are 48 SAUs and one Central Agricultural University (CAU). SAUs and ICAR institutes are also undertaking extension activities to transfer developed technologies to the cultivators. However, the target population to be covered by the available extension and education system has limitations in reaching out to all citizens in a geographically vast country like India, particularly those in rural and agricultural-based communities.

Need for Open and Distance Learning (ODL)

Agricultural knowledge and advancement in farming practices must be brought to the farming for their adoption. Open and distance learning is a proven system of education for the un-reached. Distance education is a form of education which relies on the pedagogy/andragogy, technology and instructional systems design that are effectively incorporated in delivering education to students who are not physically "*on site*". However, teachers and students can interact at a time of their own choice. The mode of interaction can be in the form of print or electronic media or technology mediated. Because of use of multi-technology, openness and flexibility, the ODL system is able to become an alternate and cost effective mode of extension education. The ODL System can provide access to extension system to reach large segment of population.

Indira Gandhi National Open University (IGNOU)

IGNOU was established in 1985 in order to provide cost-effective quality education to large sections of population, including those living in remote and far-flung areas. The University at present offers 475 programmes of study to cumulative student strength of over 2.6 million through 21 School of Studies making it the largest University in the world (Anonymous, 2011). These programmes have been created with a view to fulfilling the learners' need for acquisition of professional qualifications; continuing education and professional development in the workspace; skill improvement and self-enrichment; and certification. IGNOU is the National Resource Centre for Open and Distance Learning providing access to sustainable and learner-centric quality education by using innovative technologies and methodologies and ensuring convergence of existing systems for massive human resources required for promoting integrated national development and global understanding. The teaching-learning process is through self-instructional learning material and multimedia contents which are supported by teacher-learner and learner-learner interaction at study centers..

School of Agriculture (SoA)

The School of Agriculture was established in January 2005 at IGNOU and is mandated to improve and sustain productivity and quality of human life in rural areas through ODL System in agriculture and allied sectors. It aims at improving out-reach of agriculture education and extension for rural masses through academic, continuing and extension programmes. The School is playing important role in turning the rural unemployed youth to first rate agricultural entrepreneurs and agri-business managers of tomorrow. This will strengthen human resource base through quality education and generate trained manpower. It imparts knowledge and skills for life-long learning. The major functional areas identified by the School to cater the educational needs of the rural community are Plantation Management, Agricultural Policy, Value Added Products from Fruits and Vegetables, Dairy Technology, Meat Technology, Production of Value Added Products from Cereals, Pulses and Oilseeds, Fish Product Technology, Organic Farming, Sericulture, Water Harvesting and Management, Poultry Farming, Bee-Keeping, etc.

The programmes offered by the School of Agriculture are mostly vocational in nature, having both theory and practical components. Hands on training on preparation of different value added products; marketing and others aspects are provided. The School of Agriculture in IGNOU also conducts need based extension/ awareness / short training programmes in the areas of Dairy Farming, Value Added Products from Fruits and Vegetables, Watershed Management, Marketing Management of Horticultural Produce, Water Harvesting, Integrated Farming System, Vermicomposting, Aquaculture, Extension Methodology for Extension Functionaries, Technologies in Increasing Milk Production, Sericulture, Hygienic Meat Production and Processing, Poultry Farming, Fruit Preservation and Pig Rearing. These training are organised using the local expertise and resources.

Technological Innovations for Improving Outreach and Effectiveness of Open and Distance Education in Agriculture

There is a need for extension system which fast and efficient in disseminating the knowledge. A number of technological innovations are used in distance education with the advent of the information and communication technologies. The Information and Communication Technologies (ICT) are bringing about the knowledge revolution by empowering the resource-poor farmers with up-to-date knowledge about agricultural technologies, best practices, markets, price trends, consumer preferences, sources of finance, weather, soil-moisture conditions and the environment (Singh, 2006). Use of innovative media and Information Technology (Radio, TV, Farm information and advisory centre, Private portals, Public and private information shops, cyber kiosks, teleconferencing, on line and e-learning) to disseminate the knowledge on agriculture has become important and popular.

IGNOU has made an attempt to conduct the educational programmes / courses using ICTs. Few programmes offered by IGNOU have been converted in to online format using ICTs. ICTs supported education if designed and implemented well, can promote the acquisition of the knowledge and skills that will empower students for lifelong learning.

The extension and educational programmes of the IGNOU have multi-media support. The University has state of art facilities for production of audio/video programmes and their dissemination through radio and television. The University also uses interactive radio counselling as well as tele-conferencing to provide interactivity in the teaching-learning process. IGNOU is the nodal agency for running a 24-hour educational TV channel (*Gyan Darshan*) and a network of 40 FM Radio stations (*Gyan Vani*) exclusively devoted for education. The University has made a major breakthrough in distance education and training with the launch of an exclusive Educational Satellite, EDUSAT which has a wide reach to cover large sections of population with its two-way video conference facility.

The instruction methodology in IGNOU is more learner-oriented and the student is an active participant in the teaching and learning process. Most of the instruction is imparted through distance rather than face-to-face communication. The system of instruction followed in IGNOU includes the following components:

- ✰ Study material
- ✰ Practical training
- ✰ Face to face counselling
- ✰ TV broadcasting
- ✰ Radio broadcasting
- ✰ Teleconferencing
- ✰ Internet based delivery
- ✰ Work experience

The agricultural programmes of IGNOU are offered through Programme Study Centres (PSCs) established at conventional research and educational institutions / organizations. The theory and practical counseling (contact session) are conducted at these PSCs. The counseling sessions are conducted by the locally available resource persons. Besides face to face contact sessions, Information Communication Technologies (ICTs) based counseling are conducted. Brief descriptions of ICTs used by the IGNOU are given below:

i. *Teleconferencing*: Two-way interactive tele-counselling, tele-teaching, tele-training, tele-discussion and extended contact programme (live sessions) are conducted via satellite (EduSat) in entire country. These tele-sessions are also webcasted at www.ignou.ac.in.

ii. *Audio-video programmes*: Demonstration of skills, better clarification and enhancement of understanding of the course material are facilitated by audio and video programmes The audio and video programmes are being played at the study centers during the hours of the counselling sessions. The video programmes are telecast at National Network of the *Doordarshan* and the *Gyan Darshan*.

iii. *Interactive radio counselling*: This is done through National Network of AIR and the *Gyan Vani* stations where teacher explain the queries raised by students *via* Toll free telephone facility.

iv. *Gyan Darshan*: IGNOU run the two TV Channels (*Gyan Darshan-I and Gyan Darshan-II*) a fully digital 24-hour exclusive Educational TV Channels. The video programmes developed for various courses are broadcasted on these channels. Live sessions of teleconferencing are also broadcasted on these channels.

v. *Gyan Vani:* It is a Cooperative of Radio and is a unique low-cost, interactive medium for enhancing and supplementing the teaching-learning process by reaching out to widespread learners. Its main objective is to bridge the gap between educationally privileged and deprived. Total around 40 FM radio stations are fully operational.

vi. *Touch Screen Kiosk*: IGNOU has put a Touch Screen Kiosk at its Student Support Service Centre and its all Regional Centres throughout the country where students by entering their enrolment number can have information related to address, admission, registration, grade card, semester wise term end examination results etc.

Various improved pedagogic methods of ODL are being developed and used for the purpose of knowledge dissemination, which are the important components of extension education.

National and International Collaboration for Development of ODL System

Distance education in agriculture is being developed and adopted in various parts of the world and different models. These programmes could provide valuable information and course materials through appropriate partnerships. International agricultural research centers offer a large set of knowledge that could be used in the national systems of education. The international education systems in distance learning should be linked with national organizations in order to strengthen knowledge access to various stakeholders of agriculture. The participation of the private sector and Non-Government Organizations (NGOs) can further complement and supplement the efforts of public sector in educating farmers and rural youth.

The School of Agriculture of IGNOU is undertaking number of programmes in collaboration with various national and international organizations like Ministry of Agriculture, Ministry of Food Processing Industries, Ministry of Rural Development, Agricultural and Processed Food Products Export Development Authority (APEDA), Central Silk Board, International Network for Bamboo and Rattan (INBAR), Commonwealth of Learning (CoL) and International Food Policy Research Institutes and International Crop Research Institute for the Semi Arid Tropics (ICRISAT). ICAR institutes, State Agricultural Universities (SAUs) and other research and educational organizations are actively helping IGNOU in developing and delivering of ODL programme in agriculture sector.

Technology Mediated Open and Distance Learning

Use of Information and Communication Technology is a well recognised fact in the present era for education delivery particularly in Open and Distance learning. This development is leading towards Technology mediated learning which refers to systems of teaching and learning in which a technology other than print has a major role. There are two existing major forms of technology mediated learning: (i) stand-alone *e.g.* computer-assisted learning and computer-managed learning and (ii) conference *e.g.* for example, audio, video or computer (www.col.org). Education counselling through Radio and video conferencing are a common feature in all programmes of IGNOU. IGNOU has also started two online programmes in agriculture. In these programmes, all course contents are available online, teaching-counselling are done online, and examination are conducted online.

The School of Agriculture is implementing a National Agriculture Innovation Project on "Innovations in Technology Mediated Learning: An Institutional Capacity Building in Using Re-usable Learning Objects in Agro- Horticulture". The project will scale upto the success stories of Open and Distance Learning. The main focus of the project is on e-extension in agriculture in the form of Reusable Learning Objects (RLO).

RLO Technology is a new paradigm in Life Long Learning. RLOs are the e-extension resources available freely and openly to anyone who can use, re-use, remix, recycle and redistribute it in restricted or un-restricted manner. RLO is the smallest stand alone and an independent unit of learning designed for re-use in multiple instructional context. A RLO is a type of online instruction that provides a digital educational resource that can be scaled and shared from a central online repository in the support of instruction and learning. Each RLO supports a single learning objective. RLO can include text, audio, video, photograph, animation, table, chart, laminations that can be suitably combined to form new body of knowledge in the form of learning lesson, unit, module, course, programme, etc. Every RLOs is searchable *i.e.* it is tagged with meta-data. In addition to this RLOs are easy to update which makes easy access to updated quality teaching and learning resources for a wide range of learners.

Conclusion

There are number of similarities in philosophy and methods of distance education and extension education. Agricultural universities, colleges and research institutes have been set up to provide the extension education. The majority of peoples in agriculture sector are resource poor; they can not effort to go these conventional universities and colleges. The numbers to be covered by the extension education are too large. Therefore, the Open and Distance Learning system can help greatly in order to impart timely and cost effective education in agriculture. Advent of new information and communication technology further improving the outreach of distance learning system. School of Agriculture of IGNOU is offering a number of need based distance learning programme to fill knowledge gap of stakeholders of agriculture. The learning materials of IGNOU are supported by the various ICTs based tools like TV, radio, audio-video programmes, teleconferencing, interactive radio counselling, etc. Several government departments and ministries are also helping IGNOU in imparting agriculture education programmes. The project on e-extension in the form of Reusable Learning object is a new paradigm in lifelong learning in agriculture. It will scale up the success stories of ODL in extension education.

References

Ahammed, M.A. (2006), ICT in Indian Agriculture: ICTs catalyze the Rural and Farm Economy, www.i4d.csdms.in.

Anonymous, IGNOU Profile 2011, IGNOU, New Delhi

Balaji, V. (2006), Knowledge at Core of Strategies, *The Hindu Survey of Indian Agriculture 2006*, p. 219-221.

Hansra, B.S. and P.K. Jain. (2010). Open and Distance Learning for Sustainable and Rural and Agricultural Development, In. R.K. Samanta (Ed.), *Revamping Agricultural Education in the Context of Globalization.* (p165-176). B.R. Publishing Corporation, Delhi.

Hedge, M.R., Extension strategies of IIHR, http://www.iihr.res.in

Singh Shadev (2006), Selected Success Stories on Agricultural Information Systems, *APAARI Publication*: 2006/1.

Sulaiman, R.V. (2000), Agriculture Extension in India- The Next Step, *Policy Brief* 9, National Centre for Agricultural Economics and Policy Research, New Delhi.

http://nac.nic.in/concept per cent 20papers/Agriculture.htm

http://pib.nic.in/

www.ignou.ac.in

http://www.col.org/resources/publications/trainingresources/Pages/intro.aspx

http://en.wikipedia.org/wiki/Distance_education

Chapter 4

Emerging Agricultural Extension Models

V. Veerabhadraiah

Agricultural Extension is an instrument of change. It enables farmers to seek and adopt desirable changes in farming. Its focus is on technology improvements, leading to human resource development. Agriculture normally keeps changing, either slowly or rapidly. The role of agricultural extension is to make this process more purposeful. And, as the agricultural sector changes, the tasks of the agricultural extension will also change, while its basic functions remain essentially the same.

A Basic Model of Extension Education

Extension is a change inducing system, which enables farmers to improve their living mainly through their own efforts. By upgrading the knowledge and skills of farmers in their occupations, Extension helps them to increase their productivity, production and their incomes.

Extension education *is a system of working with farmers, their families and communities, using informal educational procedures, to relate useful, practical new knowledge to their needs and interests, thereby enabling them to make satisfactory improvements in their living (Dwarakinath, 2011).*

Farmers are interested in meeting their needs and problems in order to create more satisfying situations. The extension agency helps them in this effort. Extension systems works with local farming communities to promote improvements in farm production, family life and community living. It has access to new knowledge useful to farmers. This body of new knowledge is the starting point of all extension efforts.

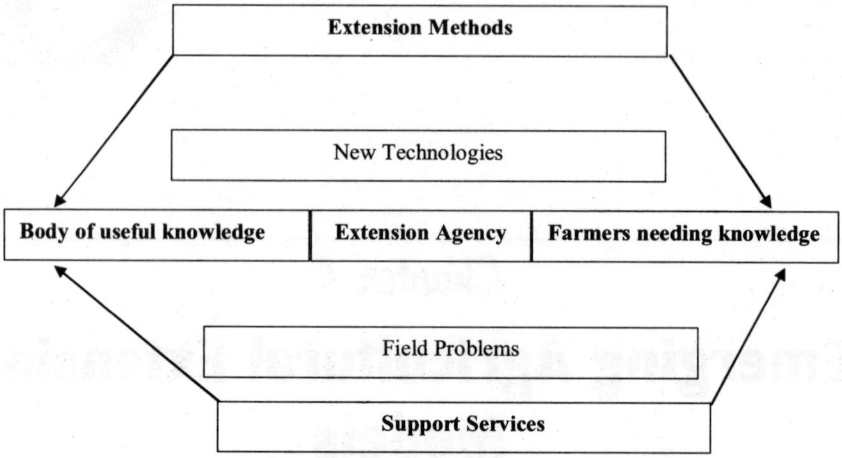

Figure 4.1: A Basic Model of Extension Education.

Components of the Model

The basic model of Extension system consists of five components. They include: (1) a body of new and useful knowledge, (2) farming communities who need the knowledge, (3) an extension agency to link the two, (4) a set of extension educational methods and (5) infrastructure and support services.

How does Extension System Operate?

The main task of extension is to make available to farmers new perceptions, practices and products for which they have a need.

- ☆ It works as a system of informal education helping people to improve their living. It remains educational in its objectives, principles and methods.
- ☆ It develops informal working relationships with individuals and groups, studies their needs and problems, develops problem solving approaches in partnership with them and gets them implemented.
- ☆ It maintains links with research agency in order to ensure continuous flow of useful information and to feed back field problems for research.
- ☆ It fosters growth of the required infrastructure facilities to support development including supplies, services and marketing.
- ☆ It encourages farmers to evaluate the changes they have made to derive lessons for future application and to derive a sense of accomplishment as a motivation for further effort.
- ☆ It encourages farmers to organize themselves into groups and associations to safe safeguard and advance their interests (Dwarakinath, 2011).

Extension Aims to

1. Change production systems and sustain the resource base
2. Increase and improve farmers' productivity and income on sustainable basis

3. Attain food security and improve livelihoods of farmers.
4. Attain higher levels of efficiency in the farm enterprise
5. Improve well being of farmers

Major trends shaping the delivering of extension services moving extension far beyond technology transfer

1. Dwindling Government budgets for extension, leading to the privatization of public services.
2. Emergence of new service providers in the context of privatization
3. Increased differentiation among client groups and issue areas (private demand driven extension for commercial farmers', public livelihoods focused extension for poor and subsistence farmers', community management support for environmental and natural resource management challenges).
4. User demand for effective and appropriate extension related to higher expectations of rural communities and farm families and their growing capacity to voice demands.
5. Globalization and liberalization of agricultural markets.
6. Increased presence of agriculture business and commercial farming.
7. Environmental imperatives
8. Climate changes.
9. Increased use of biotechnologies.
10. Access to new ICTs – increased telephone, cell phone and internet access in rural areas.

High Expectations of People for Extension

1. Efficient, demand driven extension services based on partnerships between Government and Private Sectors.
2. Farmers participating in extension programme formulations, implementation, monitoring and evaluation ensuring their needs are met.
3. Private sector including NGOs and CBOs participating in extension service delivery in response to farmers' demands.

Emerging Extension Models

Extension Education Planners throughout the world face the different challenges of being creative in their programming development efforts and responsive to the needs of rural communities and farmers. In today's world, people are continuously involved in change process either as passive elements or as active citizens, more often as mere project recipients or targets.

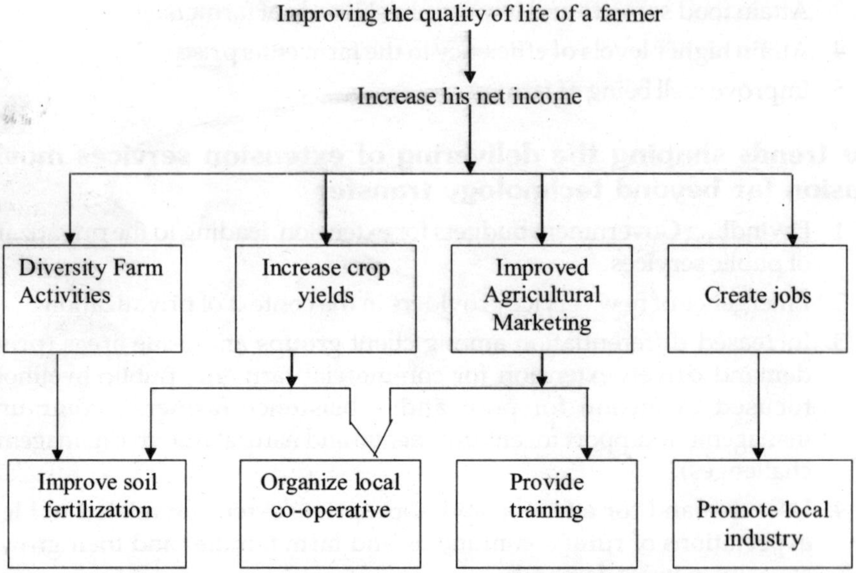

Figure 4.2: Purpose of any Emerging Extension Model.

Participatory Model: Non-Government Paradigms in Agricultural Extension

The predominant linear view of technology transfer is insufficient for addressing complex agricultural problems that exist, especially in a limited resource country like India.

The time tested models like T and V system of Extension and others are mostly top-down approaches and they are one-way paradigms. The inadequate linkages between research and extension establishments are a major weakness of development. Public sector extension systems have had limited success working with resource-poor farmers.

The NGOs (CSOs) play an increasing important role in agricultural extension in developing countries. The ability of public extension institutions is especially limited due to inadequate human resource and facilities.

Due to their good linkages with resource – poor farmers and grassroots organizations the NGOs have considerable advantage. They offer some refreshing alternative programmatic approaches compared to dominant agricultural technology transfer / extension paradigm.

Participatory Features: Facilitation and Networks

- ✩ The approach of NGOs is different from traditional extension models. They adopt participatory approach, progressive development and execution and operating through grassroots network.
- ✩ The participatory approach is significantly different from the predominant technology transfer paradigm. The farmer is not viewed as the "end

recipient" as in the case of top-down technology delivery models. Likewise extension worker is not just viewed as a "conduit" (channel) for conveying information to farmers.

- ☆ Rather than formulating a pre-packaged prescription for farmers, more effective NGOs regard farmers are valued partners in the development process at all levels of intervention need assessment, programme development, participatory technology development, implementation and evaluation. The partnership model entails giving decision making to farmers.
- ☆ For NGOs, projects are not the objective, they are only means. People have to discuss and arrive at their own solutions.
- ☆ More scope for articulation and participation of farmers. Engaging a local volunteer as promoter (may be called SAP - Sustainable Agriculture Promoter as in the case of AMEF in Karnataka, Andhra Pradesh and Tamil Nadu). The SAP should have willingness to serve people, how he interacts with other people and his respect for others.
- ☆ Linking NGOs to the Government departments and Universities.
- ☆ An innovative model of agricultural extension (agricultural department) is steadily evolving and maturing among NGOs. It has two importance characteristics – participatory at the grassroots level and proactive at the institutional level.
- ☆ The NGO development model is clearly different from the dominant agricultural extension of technology transfer model.

The change agent's role should not be *reactive* (conveying a designated technology to farmers for rapid adoption) but more *interactive*, cultivation of co-learning partnership with farmers.

Development Paradigms

Perspective Roles		*Technology Transfer Models*	*Participatory Proactive Model*
Farmer	:	Recipient	Partner
Change agents (Extension Worker)	:	Conduit (Channel)	Facilitator
Method	:	Reactive / Selling	Interactive
Researcher	:	On-station	On-farm
Management	:	Top-thinksLocal acts	Thinking and acting at all levels
Vision	:	Top-down	Shared
Approach	:	Prescription	Participation
Evaluation	:	Cost-reduction	Farmer participation

The Need for Grater Collaboration

The area of NGO-Government relationship is decisive for the strategic development and impact of NGOs. To achieve wider relevance and to gain replicability for their initiatives on large scale, the NGOs must influence government bodies both at local, state and national levels.

It is **through collaboration rather than competition,** that both public sector and NGOs can work towards more **responsive development models**. As proactive brokers of agricultural technologies, NGOs enhance change agent integrity and help build trust with farmers, thus establishing the livelihood of more interactive extension opportunities.

- ✰ The government departments stand to benefit tremendously by allowing private and voluntary efforts to take root in the society there by provide effective entry point for public sector programmes.
- ✰ When a project is given to a NGO with funds, they can begin to work immediately. The government can work with then collaboratively sharing their experience.
- ✰ As intermediaries, NGOs serve as forceful advocates of change at grassroots level. They often bridge the gap between public sector extension and resource – poor farmers.

Partnership – Extension Model

It is a measure for improving links with stakeholders. The process of agricultural information dissemination involves key actors and roles. It involves both public and private extension personnel. The actors are Government Department, private sector and NGOs, research organizations, university extension departments and farmers.

This was tried in Nigeria to facilitate adoption of banana technologies among farmers. The partnership could be sustained by identifying and selecting committed partners. The use of many organizations would promote higher adoption of new technologies among farmers.

Felicitation Model of Extension

This has emerged in recent years. It emphasizes on farmer networking to reinforce individual learning, centered with in a process which is facilitated by highly trained outside agricultural professionals (both researchers and extension workers), thus comprising Agricultural Knowledge and Intensive System (AKIS)

References

Dwarkinath, R, 2009, Reformulating the Agricultural Extension System, AMEF policy Advocacy series-4, AME Foundation, Bangalore.

Dwarakinath, R, 2011, Agricultural Development Beyond Green Revolution, Institutions of Agricultural Technologists, Bangalore.

Rolling, N. and Van de Fliert, E., 1994, Transforming Extension for sustainable Agriculture: The case of IPM in Rice is Indonesia. Agriculture and Human Values 11(2/3): 96-108.

Swansoon, B.E., 2008, Global Review of Good Agricultural Extension and Advisory Service Practices. Food and Agriculture organizations, Rome.

www.fao.org/docrep/W5830E/w5830e09.htm.

www.aiaee.org/attachments/454-mattocks-vol.1.1-9.pdf.

www.banana2008.com/cms/details/acta/879-86pdf.

Chapter 5

Technology Application in different Farm Environments

A.K. Singh[1], Lakhan Singh and Jitendra Chauhan

In India, transformation of agriculture could be realized with much needed interventions of National Agricultural Research System (NARS) in the form of various revolutions in agriculture and allied sectors. The benefits of transformational technologies were derived as they were efficiently utilized by the individual farmers in their local situations.

The commonly alleged observations quite often we hear that revolutions are confined to chosen areas, involving certain farmers, crops, livestock and commodities. The main reason for this is not the scarcity of agricultural knowledge, research findings and technologies, but lacking in translating them in to production accomplishments especially in location specific micro farming situations for the betterment of farming community. Therefore, it largely depend on understanding the various aspects in totality how the technologies are generated, assessed, refined, demonstrated and disseminated to integrate farming system as a whole.

Robert Chambers evolved the concept of Farmer First in 1987 which had insights of three broad categories of types of agriculture (Industrial, Green Revolution and the CDR or Complex, Diverse and Risk-prone); the pipeline approaches and methods of transfer of technology (TOT) for the uniform and controlled conditions of industrial and green revolution agriculture did not fit CDR conditions; farmers' practices seen as adaptive performance; the proposition that adoption by farmers is validation of a technology; the comparative advantages of farmers over scientists in innovating for complex systems; and many others. Farmer First was established as paradigmatically different from TOT, and vital for CDR agriculture. It became a movement.

Five years later, in 1992, Ian Scoones and John Thompson came with Beyond Farmer First concept. This stressed perspectives that broadened and complemented Farmer First: the pluralism of different knowledges; the recognition of knowledge as not a stock but a process; seeing farmers, extensionists, scientists and others as social actors; recognizing political dimensions and the significance of power relations; and elements of a new professionalism in agricultural science.

A workshop on Farmer First Revisited was held in 2007 which differed from the original Farmer First. In farmer First, it was mutual recognition of marginalized innovators, the solidarity of heretics, the sense of being a vanguard, of having a common commitment that could be transformative. In Farmer First Revisited it was seeing how far we had come, how many more domains than just farmer participation were relevant, and how rich the range of innovations had been. In Farmer First the focus was on the complexity and diversity of farming systems and the creativity of farmers. In Farmer First Revisited it was the complexity and diversity of domains of action and intervention and of relationships, and the co-creativity of many different actors.

In Farmer First concern was to move beyond the reductionism of production and productivity and to privilege the complex, diverse and risk-prone realities of the majority of farmers, focusing on participation on-farm with and by farmers. Now the universe itself is complex and diverse. Many aspects are multiple or multi; multiple stakeholders, multiple perspectives, multiple realities, multi-functional agriculture, multi-method approaches. There are concepts and domains that are new or new in emphasis like food systems, trade, market chains, value chains, innovation pathways and most of all innovation systems.

Now we have far many agencies other than public organizations and various kinds of public-private-farmers partnerships. Earlier farming was an adapting performance. Participatory approaches and processes were central. Now relationships and interactions are seen more clearly to have dimensions that are related to power, trust, transparency, virtual, grass root, peer, and advisory network. Partnerships like public-private, multi-stakeholder partnership etc, and collaborations like co-management, co-breeding, co-creation, co-evolution, co-development, etc. have emerged in present day context.

Different sets of support systems are emerging in many states like micro irrigation system in Gujarat, poly house technology in Uttarakhand, three tier model of Amul in Gujarat, mother dairy and fruit and vegetables marketing model by NDDB, production and export of high value fruit crops in Maharashtra have been taken up with full support from state governments and private agencies.

Since the 1990s, Farmer Field Schools (FFSs) have been utilized as a vehicle of technology application in different projects of diverse sectors. A private initiative of Mahindra Shubh Labh Services Ltd. in 2001 aimed to establish franchises of Mahindra Krishi Vihar (MKV) to provide access to inputs and machinery, credit, and advisory and field supervision services.

The micro irrigation system has been taken up as a movement in Gujarat as a tool for wise usage of resources resulting in higher agricultural productivity, thus improving the livelihood and empowering farmers to participate meaningfully in the growth of the Agriculture Sector in the State.

Uttarakhand faced with factors like undulating topography, fragile environment, fragmented and small holdings, little or no irrigation facilities, poor marketing structure has launched an ambitious project to set up cluster-based 20,000 modern polyhouses to boost production of vegetables and flowers.

The polyhouses are proposed to be constructed with an area size ranging from 100 sq meter to 200 sq meter. Through this project, the government is aiming to increase four to ten times higher productivity of cash crops. The project will also provide support for capacity building of local youth and employment of local artisans in the construction of polyhouse designs.

The Amul Model structure consists of a Dairy Cooperative Society at the village level affiliated to a Milk Union at the district level which in turn is further federated into a Milk Federation at the state level. The above three-tier structure was set up in order to delegate the various functions, milk collection is done at the Village Dairy Society, Milk Procurement and Processing at the District Milk Union and Milk and Milk Products Marketing at the State Milk Federation. This helps in eliminating not only internal competition but also ensuring that economies of scale is achieved.

Madhya Pradesh became the first state in the country and perhaps the only state in India to have a private extension policy. The policy states that the private extension would aim for cost reduction, improving the efficiency of extension system and inculcating accountability in extension services.

Considering changed socio-economic perspective, climate variation, agro-climatic situations, and other institutional factors, the technology needs to be generated accordingly. Different technology application models again need to be developed to suit the micro environments. A large network of research, social, and developmental organizations are being working at field level, need close convergence for desired outcome.

Chapter 6

Participatory Approaches for Women in Agriculture

U.S. Gautam, Alka Singh and S.R.K. Singh

Introduction

The women play a significant and crucial role in agricultural development and allied fields including in the main crop production, livestock production, horticulture, post harvest operations, agro/ social forestry, fisheries, etc. (Prasad and Singh 1992). The nature and extent of women's involvement in agriculture, no doubt, varies greatly from region to region. Even within a region, their involvement varies widely among different ecological sub-zones, farming systems, castes, classes and stages in the family cycle. But regardless of these variations, there is hardly any activity in agricultural production, except ploughing in which women are not actively involved. In some of the farm activities like processing and storage, women predominate so strongly that man workers are numerically insignificant. (Aggarwal 2003) Studies on women in agriculture conducted in India and other developing and under developed countries all point to the conclusion that women contribute far more to agricultural production than has generally been acknowledged.

Therefore, a great need has been felt with consideration of substantiality and equity for developing agricultural research and development programmes which will reflect the role of women as farmers, thus, enabling the women to raise their standard of living and contributing their skills in nation building process. Hence, Indian Council of Agricultural Research (ICAR) has established a network of Krishi Vigyan Kendras (KVK), which was rest upon a solid base of transfer of technology from laboratory to farmer's field with respect to Agriculture, Horticulture, Animal Husbandry, Floriculture, Bee keeping, Mushroom Cultivation, Broiler Farming and allied subjects to the farmers, farm women, rural youth, school drop outs and extension functionaries to increase their knowledge, enhance their skill and change their attitude in concerned sectors.

As per the recommendations of Mohan Singh Mehta Committee during 1974 the first KVK was established in Pondicherry and by the end of IX Plan, the numbers of KVKs were 276, now 575 KVK are running in India. Gradually working guidelines are prepared to make the K.V.K. as the light house for the rural people.

Role of women in agriculture is hugely emphasized in KVK. One of the principal thrust area of KVK is to make farm women participation more in agriculture and allied aspects. KVK is exploring the possibility of entrepreneurship development for rural women through agriculture.

The KVK today has sufficient resources to impart training skills for not only the farmers but also the rural youth. The training schedule typically incorporates the existing needs and problems of the farmers for making a positive impact. The trainings are conducted both at the on-campus and off-campus locations. It invariably emphasizes on providing both the short term as well as long duration courses specifically to impart practical orientation to these courses. The KVK has started the instructional units for not only imparting the skills but also for providing the critical inputs as per the demand and need of the farmers.

Mandate of KVK

The mandate of KVK is technology assessment, refinement and demonstration of technology/products. The activities of KVK include on-farm testing to identify the location specificity of agricultural technologies under various farming systems, frontline demonstrations to establish its production potentials on the farmer's field, training of farmers to update their knowledge and skills in modern agricultural technologies, and training of extension personnel to orient them in the frontier areas of technology development. As recommended by the committee on Independent Evaluation and Impact Assessment of KVK and QRT, the fourth activity of the KVKs will be to work as resource and knowledge centre of agricultural technology for supporting initiatives of public, private and voluntary sector for improving the agricultural economy of the district.

Scope for Women in Agriculture

Since time immemorial women have significant contribution in almost every aspects of our society. Besides having an anchoring role in the management of their families, women equally participate in different economic activities like crop production, vegetable cultivation, fruit raising, livestock management, pisciculture and other miscellaneous activities such as mushroom cultivation, leaf plate making, bamboo works, badi and papad making etc.

Meeting world food needs in the year 2025 will depend even more than it does now on the capabilities and resources of women. Women are responsible for generating food security for their families in many developing countries, particularly in India. Women not only process, purchase, and prepare food, but they also play a significant role in national agricultural production, producing both food and cash crops. Population growth, urbanization, and the limited potential for increasing production through the expansion of cultivated area imply that, for food needs to be met in the future, yields will have to increase. Agricultural research continues to develop new varieties with higher yields and increased tolerance to unfavorable environmental conditions, but an untapped source of productivity gains could lie in addressing gender disparities in agriculture. The women play in maintaining the three pillars of food security-food production, food access, and food utilization-

Women as Food Producers

The sustainable production of food is the first pillar of food security. Millions of women work as farmers, farm workers, and natural resource managers. In doing so they contribute to national agricultural output, maintenance of the environment, and family food security.

Women as Gatekeepers

The second pillar of food security is making sure that household members, particularly the children, receive an adequate share of the food that is potentially available. Women act as the "gatekeepers" of their households' food security through the allocation of their own time and income.

Evidence suggests that men spend a higher proportion of their incremental income on goods for their personal consumption. By contrast, women are more likely to purchase goods for their children and for general household consumption.

The third pillar of food security-Food Utilization-means ensuring that the food consumed contributes to good physical and cognitive development. This entails the provision of "care," namely, paying adequate time and attention to meeting the physical, mental, and social needs of growing children and other household members. Care affects food security in two broad ways: first, through feeding practices such as breast-feeding and the preparation of nutritious foods for weaned infants and others in the family, and second, through health and hygiene practices such as the bathing of children and the washing of hands before food preparation. These caring behaviors are time-intensive. The time constraint is especially acute for female heads of households.

Multi-Dimensional Role of Women

1. Agriculture: Sowing, transplanting, weeding, irrigation, fertilizer application, plant protection, harvesting, winnowing, storing etc.
2. Domestic: Cooking, child rearing, water collection, fuel wood gathering, household maintenance etc.
3. Allied Activities: Cattle management, fodder collection, milking etc.

Women in Decision Making in Agriculture

Women's role in agricultural operations, animal husbandry and other economically productive activities is very significant. They contribute about 60-70 per cent of the labour required for these activities thus playing a pivotal role in sustaining economy. The decision making process is an important segment of every household because it makes implementation of a plan or programme quite easy. In rural areas of the country, both husband and wife are jointly responsible for making decisions on matters like family obligations, specific housing charges and purchase of household articles. However, women's suggestions are not given due consideration in the decisions pertaining to agricultural sector and important family matters. It is because the majority of women are illiterate, has little time to know about the latest techniques of farming and restricted mobility due to several cultural taboos. At times, even the suggestions of knowledgeable rural women are ignored or are not taken seriously because men consider it disgraceful to accept the decision of women. This is because traditionally men have been major lawmakers of society. Many policies and decisions neglect women and undermine their abilities and roles. The programmes should be developed exclusively for women; to build leadership skills for managing agricultural community based development activities.

Storage

Storage of food crops is critical to achieving food security on both a national and local level by easing price variations and assuring more stable supplies. In India, storage facilities are inadequate and ineffective. It is estimated that one-fourth of all food produced is lost due to spoilage, insects and rodents. Women play an important role in food storage and handling, and as such could benefit from improved methods and facilities. Improvements in traditional systems of storage and food handling could make a major contribution to food security in terms of stabilizing supplies and helping to maintain the nutritional value of stored and processed foods.

Food Preservation

Women perform majority of the work in food processing in developing countries. Women are responsible for supplying their families with food and care, they often have special knowledge of the value and diverse use of plants for nutrition, health and income. Consequently, they are frequently the preservers of traditional knowledge of indigenous plants. Women process and preserve the fruit and vegetable produce from their home gardens and from the forests. Women are universally responsible for preparing food for their households and thus for the nutritional well-being of its members.

Women in Crop Production

Women's participation in crop production has increased as a result of male migration to urban areas to seek employment. Women are active in all aspects of crop production, expect ploughing in which women are not actively involved. Women are involved in sowing, weeding, plant protection, maintenance, harvesting, food preparation, storage, preservation, treatment and marketing of crops.

Women in Dairy Sector

Rural women, who constitute nearly 77 per cent of the total female population of the country, play an important role in agriculture and animal husbandry besides the household responsibilities. Agarwal (1987) reported that the involvement and participation of rural women were more in animal husbandry than in agriculture. It is true that a nucleus of dairy unit more or less exists as a traditional component in every rural household as a source of draught power or nutritional substance for the family, or a secondary source of income. Venkatachalam (1993) and Sethi (1991) expressed that almost 80-90 per cent of the total work related to cattle care is performed by women. In many places the entire management of livestock, starting from cutting, collection, carrying and chaffing of fodder to feeding and milking, preparation of milk products, cleaning of cattle shed, collection of cowdung to the manure pits, preparation of cowdung cakes and their storage was done by women (Achanta, 1982). The short duration training programmes on dairy practices would be useful to the farm women in acquiring the latest technical knowledge and skills with regard to dairying.

Women in Poultry Production

Poultry production is an occupation that supports the livelihoods of many women, especially the tribal and rural poor in India. Poultry act as financial reserves in crisis situations, and thus reduce vulnerability. The majority of the rural poor still depend on agricultural labour, which does not offer a year round earning opportunity, and livestock rearing supplements their income. Poultry is a source of low cost, high quality food for the family and small cash for women through sale of birds and eggs. Most of the procurement and sale of birds and eggs are done within the village and in local weekly markets.

Drudgery Reduction

Women in agricultural communities work for longer hours compared to men and due to their heavy work load, women often find it difficult to attend training and other extension programmes. Releasing time from these heavy workloads is essential in order to get meaningful participation of women in agricultural technologies. There is a need to understand the daily and seasonal workloads of women farmers and accordingly suggest her for drudgery reducing improved tools and techniques. Promote sowing tools, paddy threshers, winnowers, sprayers, weeder, harvesting tools, cleaner, grader, parboiling units, maize shellers, dal making machines etc.

Role of SHGs in Agriculture Development

The importance of organising women into SHGs is an accepted method of empowerment. SHGs are an effective means through which women can build self-reliance, solidarity and confidence, as well as contribute to the overall development of their families and societies. Belonging to a group provides women with a sense of identity, status and security, and also provides them access to credit and institutional support services which they otherwise would not have had as individuals. SHGs have bargaining power and the ability to make major decisions for their own development. SHGs need skill to group formation, conflict resolution, leadership development etc, and KVK act as facilitators to help in group formation, training, entrepreneurship development etc. also offer the guidance on rapport building, analysis of social structure, livelihoods and local resources, group formation, group planning, participatory monitoring and participatory evaluation.

Selection of Enterprise

Many of the women trained under different programmes have initiated new enterprises as individuals and as groups and could increase their income substantially from these enterprises. But it is the selection of the right kind of enterprises that is critical. The choice of enterprise depends primarily on the time available with women, their risk taking capacity, cost of inputs, cash flow, markets etc. The capacity of the women groups to understand the likely outcomes of different enterprises and to help them take a decision on what enterprise would be suitable for them, could be developed through appropriate training by KVK.

The woman farmers require activities that are not too time consuming, not drudgeries and will allow them to generate income. With this in mind, the most common income generation activities are listed below:

- Purchase/leasing of agricultural land
- Improved seeds, bullocks, implements and other inputs for crop production
- Purchase of dairy cows, goats and chicks
- Tailoring
- Paddy processing
- Rope making
- Fish-net production and repair
- Mat weaving
- Production of processed milk products
- Pottery
- Vermi-composting
- Nursery raising
- Bee-keeping
- Food processing (pickle and papad production and sale, curry and spice powder units)
- Leaf cup production
- Mushroom production
- Cultivation of crops
- Garment production
- Pasture development
- Soap powder
- Agarbatti making
- Operation of flour mill
- Production of greeting cards

- ✰ Consumer store and fair price shops
- ✰ Selling of cloth, sarees and bangle
- ✰ Embroidery unit
- ✰ Stationery production and sale

Policy and Strategies

Women's participation in agriculture varies widely from different ecological sub-zones, farming systems, castes, classes etc in India. The involvement of women in crop production varies according to the type of crop grown and the cropping systems and the socio-economic status of the family. There is also a need to make distinctions between the involvement of women as agricultural labourers and involvement of women in agricultural operations on their own farm. In relatively prosperous areas where land holdings are large and most of the agricultural operations are mechanised, women play only a marginal role. The women from poorer families work as agricultural labourers irrespective of the community to which they belong. Female labour participation rates are noticeably higher among scheduled caste and scheduled tribal communities (Prameela, 1990). Even though landless women do not have own land, their role in agriculture is important as they are the ones involved in 85 per cent of the agricultural operations in India. Landless women also lease in land for cultivation. Keeping milch animals, small ruminants and backyard poultry is an important source of income for poor farm families and agricultural labourers. Thus, training programmes for women in agriculture should not ignore this group on the basis that they do not have land and do not belong to the category of small and marginal farmers. The problems of tribal women are different from other rural women and they need a totally different approach to overcome some of these constraints.

It is important to recognise the various categories of women farmers that exist and their needs in the agriculture sphere and from there to develop appropriate strategies to assist them *e.g.* whilst in some parts of India require trained women to reach women farmers, others may require trained women as motivators only, and other areas may need only the agricultural officers to be trained on women's issues to disseminate technology to women. Women must be empowered by enhancing their awareness, knowledge, skills and technology. Planners, social welfare agencies and women's development organisation should provide a rightful place to women. It is necessary to expand networking of support services so that women are freed from some of their gender imposed roles.

Technologies which are labour saving, drudgery reducing, income generating and productivity increasing should be given wide publicity and their use encouraged through pragmatic extension.

Scenario by 2025

- ✰ The farm women will have farm management skills and capable of taking even complex decisions like shifting from subsistence farming to diversified agriculture, with stand competition from market forces improvement in work or farm efficiency etc.

- ✰ Suitable agricultural interventions will enhance production of quality food grains, vegetables and fruits. This will help in improving per capita availability of food, which will, in turn, impart food and nutritional security.
- ✰ Improved access to credit, establishment of milk cooperative and marketing will enhance the managerial abilities of farm women in dairying.
- ✰ Women will act as "vaccinators" and as manufacturers in low cost dairy and poultry feed formulations at the village level.
- ✰ Women will play a significant role in livestock management and production and will be the member of dairy cooperatives.
- ✰ Availability of time saving, work ease and efficient effective implements and tools will ease out the drudgery and enhance the working efficiency of farm women and this will leave much time for other work and income generating enterprises.
- ✰ Greater awareness will bring in reduction in malnutrition in rural areas through fortification of conventional foods with nutrient rich produce such as soybean.
- ✰ Suitable gender-specific technologies will empower the farm women.
- ✰ Availability of safe and pure drinking water will reduce the incidence of diseases which leads to malnutrition, save money and increase the working efficiency.
- ✰ Reduce losses of produce by improved storage structures and technologies will raise income of the farm women.
- ✰ Lastly, it is expected that empowerment of rural women will also facilitate in achieving the Vision 2025 of the country in terms of better livelihood support, poverty reduction, building food and nutritional security for the family and a gender balanced healthy and prosperous society.

References

1. Achanta, I.D. (1982). 'Role of women in agricultural developments, Kurukshetra, vol.31(2), pp. 15-17.
2. Agarwal, Hema, (1987). 'A study on contribution of farm women in animal husbandry enterprise'. An unpublished M.Sc. thesis submitted to IVRI, Izatnagar.
3. Aggarwal, M. (2003). "Economic participation of rural women in agriculture" in Economic empowerment of rural women in India, edited by Gopal Singh, 2003, RBSA Publications, Jaipur, Rajasthan.
4. Prameela, V. (1990). A study on knowledge and participation of rural women in agricultural operations with respect to paddy crop in Kurnool district of Andhra Pradesh, M.Sc Ag.Thesis (unpublished) submitted to the University of Agricultural Sciences, Bangalore.
5. Prasad, C. and Singh, R.P. (1992). Farm women: "A Precious Resources" in Women in Agriculture, Vol.2, Education, Training and Development edited by R.K. Punia, 1992, Northern Book Centre, Ansari road, New Delhi.

6. Sethi, R.H. (1991). 'Women in agriculture'. Rawant publications, New Delhi, pp. 55-90.
7. Venkatachalam, S. (1983). 'Training of rural-Must for sound dairy extension programme', Tamil Nadu Journal of Cooperation, vol : 74 (8-9), p.475.

Chapter 7

Human Resource Development for Sustainable Agriculture and Rural Development under Present Climate Change

B.K. Narayana Swamy

Climate change is a natural process and the earth was subjected to many changes since its formation. Just as the weather changes from day to day, the climate changes over decades, centuries and millennia. However, sustainable agriculture and rural development and climate are mutually dependent. There is a need to understand the effect of climate change on sustainable agriculture and rural development sector at Global level from the view point of providing food to growing population. The increase in temperature has significantly led to a change in the sustainable agricultural zones and shift in the growing seasons. On the other hand the change in the rainfall pattern is the serious threat to the sustainable agriculture and rural development, which in turn affects the economy and food security. The several research and evaluation studies conducted revealed that utilization of human resource is far from satisfactory in different parts of the world. Five scientific revolutions have emerged under changing global climate in the world during this millennium. Human Resource Development (HRD) is essential for adoption of emerging scientific revolutions in changing climate for Sustainable Agriculture and Rural Development. Climate change, energy and Human Resource play a pivotal role in areas of productivity and quality. Further, lack of attention to the nonliving factors may result in reduction of profitability to some extent but ignoring the human resource can prove to be disastrous. However, Narayana's Innovation Attributes Lotus Model explains four attributes with seven sub components under each for appropriateness of technologies to adopt like Relative advantage, Compatibility, Practibility and Complexity in the emerging global climate change for Sustainable Agriculture and Rural Development. Further, Narayana's Wheel Model explains management services for Technical advice, Production inputs, Credit support, Marketing service, specialized service, Crop insurance and Programme subsidy are very essential for different sustainable agricultural programmes at local level to increase agricultural production and rural development.

Introduction

Climate change is a natural process and the Earth was subjected to many changes since its formation. Just as the weather changes from day to day, the climate changes over decades, centuries and millennia. But recently, the changes have become more apparent. Historical evidences suggest that the world has warmed more rapidly in the past 100 years than ever before. Climate change is one of the biggest challenges facing the world today. The problem of human induced climate change first came into force and drew the attention of the scientists and policy makers when Inter governmental Panel on Climate Change was established. The effects of global climate change are many folds and there is need to create awareness about its impact on various sectors of economy. There is an urgent need to take all necessary steps by both developed and developing countries to reduce greenhouse gases that are being let out into atmosphere. However, sustainable agriculture and rural development and climate are mutually dependent. There is a need to understand the effect of climate change on sustainable agricultural and rural development sector both at Global and as well as at regional level especially from the point of view of providing food to vulnerable section of the population. Changing climatic conditions can have the big effect on our life and our environment. The change in weather conditions can be best observed through the extreme rise in temperature, melting of glaciers and sudden rise in sea level. These changes are causing serious problems to humans and other forms of life. Sustainable agriculture and rural development in India and entire world is mostly dependent on the persisting weather conditions. The alteration in Global warming has dramatically affected sustainable agriculture and its productivity. The increase in temperature has significantly led to a change in the sustainable agricultural zones and shift in the growing seasons. On the other hand the change in the rainfall pattern is the serious threat to the sustainable agriculture and rural development, which in turn affects the country's economy and food security. The several studies conducted in different parts of the country indicated that the overall loss in the crop production in the country in the last few years due to the anticipated rise in the temperature. It is expected that in the near future India is going to face the challenges that includes unwanted pressure from the growing population, and changing scenario of world trade in sustainable agriculture. The poor are likely to be hit hardest by climate change, and that capacity to respond to climate change is lowest in developing countries and among the poorest people in those countries. It seems clear that vulnerability to climate change is closely related to poverty, as the poor are least able to respond to climatic stimuli. Furthermore, certain regions of the world are more severely affected by the effects of climate change than others. Generally speaking, vulnerability and adaptation to climate change are urgent issues among many developing countries. For this reason, there exist provisions in the United Nations Framework Convention on Climate Change (UNFCCC) to assist those countries that are thought to be most vulnerable and least able to adapt. Within the context of the Climate Change Knowledge Network (CCKN), a project on the impacts of economic changes and climate change on India's sustainable agricultural and rural development sector is being pursued jointly by the International Institute for Sustainable Development (IISD), the Centre for International Climate and Environmental Research (CICERO) and the Tata Energy Research Institute (TERI).

Strengths and Current Issues of Agriculture and Rural Development

The Strengths of Indian agriculture are like (1)Strong strides made in increasing the production in the past 50 years, mainly due to adoption of HYVs and other technological developments (2) Subsistence agriculture with small land holdings and skewed distribution of land (3)Wide variation in regional productivities (4) Majority still depend on rainfed agriculture (5) Frequently affected by extreme weather events such as droughts and cyclones (6) Significant proportion of population still under poverty, mal-nutrition and chronic hunger. The Current Issues in Agriculture are (1) Overproduction in short-term, yet food insecurity for a large population (2) Stagnation/decline in yields (3) Diversification (4) Quality and quantity of water resources (5) Increasing production cost and decrease in the price of commodity. Further, recent research studies suggested that there is a possibility that this gradual global warming could lead to a relatively abrupt slowing of the ocean's thermo line conveyor, which could lead to harsher winter weather conditions, sharply reduced soil moisture and more intense winds in certain regions that currently provide a significant fraction of the world's food production. With inadequate preparation, the result could be a significant drop in the human carrying capacity of the earth's environment. According to the National Academy of Sciences, the earth's surface temperature has risen by about 1 degree Fahrenheit in the past century, with accelerated warming during the past two decades. There is new and stronger evidence that most of the warming over the last 50 years is attributed to human activities that have altered the chemical composition of the atmosphere. The annual and monsoon rainfall has lot of variability and also the food production is mainly affected by the rainfall. This indicates that the production has been decreased during the low rainfall years. The rainfall during the season is most important and has lot of influence on the production instead of rainfall trend.

Interventions to Overcome Climate Impact on Sustainable Agriculture

The predominant interventions to overcome climate related impacts in rainfed areas include (1) Soil and water conservation practices (2) Agronomic interventions (3) Nutrient management practices (4) Livestock based interventions (5) Development of alternate land use plans These interventions have a role to play in all agro-eco systems except that their order of priority changes, which basically depends on rainfall, status of natural resources like soil, water etc. Crop based interventions need to be planned based on amount and distribution of rainfall, availability and further augmentation of water resources with watershed programme. Various soil and water conservation measures for rainfed agriculture include (1) In-situ measures for rainwater management in rainfall areas. (2) Off season land treatment: (3) Conservation furrows (4) Ridges and furrows system in cotton (5) Cover cropping (6) Micro catchments for tree systems Medium term measures rain water management (7) Stone and vegetative field bunds for soil and water conservation (8) Graded line bund helps in efficient drainage. (9)Trench cum bund for soil and water conservation Long term measures for rain water management like (1) Water harvesting (2) Contour trenching for runoff collection (3) On-farm reservoirs (4) Ground water recharges structure (percolation tanks). (5) Recharge through defunct wells. These rainwater management strategies have good scope for adoption. The increasing probability of floods and droughts and other uncertainties in climate may seriously increase the vulnerability of resource-poor farmers to global climate change. In view of these climatic changes and the uncertainties in future sustainable agricultural technologies and trade scenarios, it will be very useful to have an early warning system of environmental changes and their spatial and temporal magnitude. Such a system could help in determining programmes to the potential food insecure areas and communities. Modern tools of information technology could greatly facilitate this through HRD.

Human Resource to Use Modern Tools for Rural Development

The development of human resource is becoming a matter of prime concern for increasing sustainable agricultural production. It is the professional who create circumstances that can help in making things happen. Human Resource Development (HRD) is the process of enabling people to make things happen. India is liberalizing and globalizing the economy in changing climate. This has created unprecedent upsurge in the demand for accurate information. The several research and evaluation studies conducted in India revealed that utilization of human resource is far from satisfactory in different parts of the country. Five Scientific Revolutions have emerged under changing global climate in the world during this millennium like (1) Genetic Engineering and ability to use this understanding to develop new process and products. (2)Eco-technology and blending of best traditional knowledge with frontier technologies (3) Information technology and its rapid growth in the systematic assimilation and timely dissemination to the concerned. (4) Motivation techniques for efficient utilization of available technologies by ultimate users. (5) Appropriate policies to technology development, technology dissemination and technology utilization. Hence, HRD is essential for adoption of emerging scientific revolutions in changing climate. The professional competencies include knowledge, skill, attitude and values. The professional competency is required for application of modern sustainable Agricultural Production system under current agrarian issues. The competencies could enable professional to act and improve for more alternatives and increase choices. Of all the factors of sustainable agricultural production man has the highest priority and is the most significant factor of production and plays a pivotal role in areas of productivity and quality. Further, lack of attention to the nonliving factors may result in reduction of profitability to some extent but ignoring the human resource can prove to be disastrous. Hence, there is need for HRD to increase sustainable agricultural production under changing global climate. The how aspect is explained as follows

1. Need for Appropriate Technology Development to Increase Sustainable Agricultural Production and Rural Development

The several research and evaluation studies conducted in India revealed that utilization of modern sustainable agricultural technologies is far from satisfactory in different parts of the country under current agrarian issues. However, Narayana's Innovation Attributes Lotus Model (Figure 7.1) explains four attributes with seven sub components under each for appropriateness of technologies to adopt like Relative advantage, Compatibility, Practibility and Complexity in the emerging global order. (A)Relative Advantage: Is the degree to which an innovation is superior to the idea it supersedes. It can be explained with seven sub items like. (1)Market feasibility: as the extent of market demand for the product derived out of innovation and also the extent of scope for marketing product. (2). Market stability: as the consistency of market price and demand of product derived out of Innovation. (3). Cost: is of two types, initial cost and continuing cost. Initial cost represents the capital investment required for adoption of innovation. Further, the cash or inputs required for subsequent years use of innovation is termed as continuing cost. (4) Net Profit: as the quantum of monetary benefit obtained by an individual through adoption of innovation. (5) Profit consistency: denotes the regularity of net returns obtained by an individual or group of individuals over a period of time by adoption of an innovation. (6) Utility potential: as the degree to which the multiple use potential of an innovation to an individual or group of individuals through adoption of innovations. (7) Time saving: Indicates the best efficiency of an innovation in terms of saving time in different aspects. (B) Compatibility: an innovation is consistent with past experiences, existing values, and future plans of the adopters of technologies. Compatibility is divided into seven sub groups like (1) Cultural compatibility: an innovation is consistent with the values and norms of the society. (2) Social compatibility: denotes prestige gain or esteem by individual in the society through adoption of an innovation. (3) Physical compatibility: an innovation is consistent and should fit into the needs and interests of the adopters. (4) Psychological compatibility: Is innovation usefulness as perceived by the members of social system. (5) Situational compatibility: Denotes consistency and harmony of the innovation with previous practices followed by adopters. (6) Relational compatibility: an innovation can be adopted independently by the adopters just like other practices. (7) Anticipated compatibility: An innovation should be consistent with the future ideas of the adopter over a period of time. (C) Practibility: an innovation can be easily communicated, tested, demonstrated and practiced. (1)Point of origin: Indicates the credibility of the source from where the innovation originated. (2) Access to advice: For implementation of innovation, its extent of availability of original and detailed information for guidance and clearing doubts that arise while implementing it. (3) Visibility: the results of an innovation are visible. (4) Trialability: Is the degree to which new idea can be tried on a small scale. (5) Mastery: Is the practice of an innovation could be learned or mastered in a short period of time. (6) Demonstrability: an innovation can be demonstrated to members of social system easily. (7) Communicability: Is the information about the new idea can be diffused to members

Narayana's Innovation Attributes

Relative Advantage	**Practibility**
1. Market feasibility	1. Point of origin
2. Market stability	2. Access to advise
3. Cost	3. Visibility
4. Net profit	4. Trialability
5. Profit consistency	5. Mastery
6. Utility potential	6. Demonstrability
7. Time saving	7. Communicability
Compatibility	**Complexity**
1. Cultural	1. Failure probability
2. Social	2. Discomfort savings
3. Physical	3. Resource complexity
4. Psychological	4. Reversibility
5. Situational	5. Work efficiency
6. Relational	6. Cognitive complexity
7. Anticipated	7. Applications complexity

Figure 7.1: Appropriate Attributes of Modern Technologies are Show on the Lotus.

of the social system easily and speedily. (D) Complexity: an innovation is relatively difficult to understand and use. (1) Failure probability: an innovation chances of failure and uncertainity of results after its adoption. (2) Discomfort saving: Represents avoidance of physical discomfort may be derived by adoption of an innovation. (3) Resource complexity: difficult in getting the necessary inputs and other resources for the application of an innovation. (4) Reversibility: degree of ease with which the innovation can be replaced in case of its failure. (5) Work efficiency: the adoption of new idea saves labour or increase the available labour efficiency. (6) Cognitive complexity: an extent of relative difficult in understanding an innovation. (7) Application complexity: relative difficulty of an innovations use and application on the farm. The appropriate technology developed should make use of sustainable Agricultural information system for proper dissemination. This demand for HRD.

2. Management Services for Eco-friendly Management Practices

Management services to small poor farmers are an emerging concept cited in recent literature. However, the Narayana's Wheel Model (Figure 7.2) of management services to small farmers explains that small farmers obtain assistance in getting operational activities required like Technical advice, Production inputs, Credit support, Marketing services, Crop/livestock insurance, specialized services and Programme subsidy (TICMSIS) for different eco friendly management practices programmes at local level. (1)Technical Advice: Many times several small farmers fail to adopt recommended eco friendly management practices due to lack of technical assistance at local level. If assistance is provided small farmers adopt all recommended

NARAYAN'S WHEEL MODEL OF MANAGEMENT SERVICES FOR SUSTAINABLE LIVELIHOOD SECURITY

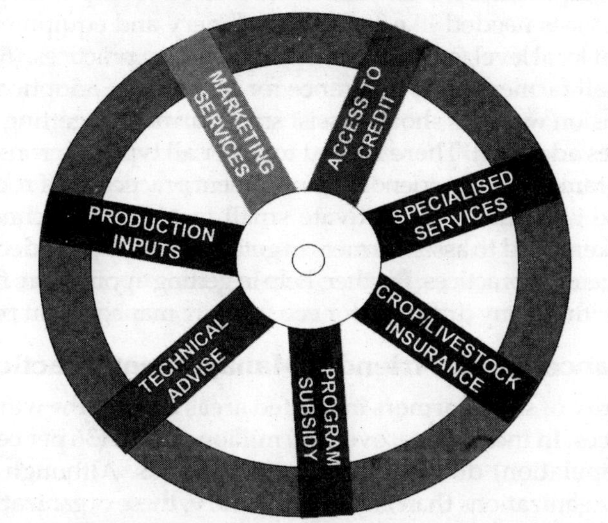

Provided management services to Seven Orchestrated, Concerted, Comprehensible Operational Activities in the agricultural system are shown on the Wheel

Figure 7.2

eco friendly management practices. Hence, they should get assistance to see demonstrations established in the locality to understand the importance of technology promoted and technical information through village meetings/training programme/distribution of literature etc. Promoters need to motivate them in respect of superiority of modern eco friendly management practices by exposing them to the economic aspects. Further, they should be helped to know the techniques at various stages of eco friendly management practices. (2) Production Inputs: Extension Workers should help small farmers in finding out the requirements of inputs and help to obtain these inputs at local level sale point, if not available at local level sale point, they should arrange to get from the nearest sale point for eco friendly management practices. (3) Marketing Services: Market for the produce is an important factor as many small farmers fail in marketing because of lack of knowledge. Hence, extension workers need to provide marketing assistance needed for small farmers in different aspects like grading, processing, storing, transport to market, pledge finance, market price information etc to avoid distress selling and middle man in marketing for eco friendly management practices. (4) Credit Support: Credit is the most important input required for technology adoption. Many a times small farmers were not able to adopt the recommended eco friendly management practices because of financial constraints. If timely finance is provided small farmers will adopt the recommended technologies and increase the production. Extension workers should help farmers in finding out the requirements of credit and in getting local application, filling, processing and sanction without any difficulty for eco friendly management practices. (5) Specialized Services: Many times small farmers feel difficult to carryout operations due to lack of equipments and they stop the particular operation or carry out in their own way with the available equipment. So extension workers need to help rural people in getting specialized services needed like getting machinery and equipment for the use of small farmers at local level for eco friendly management practices. (6) Crop/livestock Insurance: Small farmers need insurance for technology adoption to escape from risk. The extension workers should assist small farmers in getting insurance cover for technologies adoption. There is need to cover all types of crops/animals under insurance programme for eco friendly management practices. (7)Programme Subsidy: This assistance is required to motivate small farmers for technology adoption. Extension workers need to assist farmers in getting subsidy provided to different eco-friendly management practices. Further, help in getting application, filling, processing and sanction without any difficulty for eco friendly management practices.

3. Microfinance to Eco-friendly Management Practices

The majority of small farmers in rainfed areas around the world lack access to financial services. In India alone, over 200 million people (36 per cent of the rainfed agriculture population) do not have access to banks. Although India has more microfinance organizations than any other country, these organizations only reach a small percentage of needy households. The rest have no alternative than going to the local moneylenders whose exorbitant interest rates reinforce the indebtedness that contributes to lifetime poverty. Microfinance as the financial service for the small borrowers in rainfed areas and resource poor regions for their economic independence and emancipation. Microfinance as the entire range of financial services rendered to the poor including skill up gradation entrepreneurial development that would enable them to overcome poverty. Hence, microfinance has the provision of financial service to low income clients including self employed for their livelihood security in rainfed India. Microfinance is being looked upon, by all the concerned as an instrument of poverty alleviation. In India major rural development projects are being funded by the leading agencies like the World Bank, International Fund for Agricultural Development, Asian Development Bank and African Development Bank are stressing upon the need for innovative reforms in credit delivery. However, microfinance is different from micro credit. The micro credit emphasis on loans but micro finance includes support services where we open channels for thrift, market assistance, technical assistance, capacity building, insurance, social and cultural programmes. The major institutional initiatives include the bank linkage programme under the overall guidelines and supervision of the NABARD. The RBI has come out with directives on various aspects of microfinance provision for eco friendly management practices. Hence, there is need to understand the concept and importance of microfinance for eco friendly management practices of small farmers. According to Asian Development Bank, one of the biggest donors for micro-finance, provision of financial services, such as deposits, loans, payment services, money transfers and insurance to poor and their micro-enterprises are broadly called 'micro-financing'. The term micro-finance came into greater currency since the early 1990s and has largely supplemented the term 'micro-credit'. Further, micro finance play important role eco friendly management practices of small farmers in rural areas. Microfinance is providing financial services, savings and credit to small farmers who do not have access to formal financial institutions and considered world over as an important tool in poverty alleviation. The results of the study reported that micro finance programmes are unique, in that they foster the participation of the poor in the process of economic growth by creating employment opportunities. It helps in increasing the access of the small farmers to income generating assets and raising the productivity of these assets both physical and human. It enables the small farmers to manage risk better. In periods of change, it helps to mitigate the transitional costs of the adjustment process for most vulnerable groups. Its subsidized interest rates. It administered loan allocating and targeted credit programmes were designed to displace informal sources of financial services for eco friendly management practices.

4. Capacity Building among Professionals for Adoption of Modern Technologies to Increase Sustainable Agricultural Production

Under the current agrarian issues capacity building among professionals is too complex phenomenon to be explained by a single factor. However, Narayana's Wheel Model (Figure 7.3) explains combination of seven components for capacity building among professionals *viz.* Innovativeness, Decision making ability, Achievement motivation, Information seeking ability, Risk taking ability, Coordinating ability and Leadership ability. The combined contribution of the above seven factors to an individual behavior is being expressed in terms of capacity building among professionals, so far attention given is limited.(1) Innovativeness: Considered as socio-psychological orientation closely associated with change, adopting new ideas and practices. An individual adopts new ideas relatively earlier than others in his

Figure 7.3: Seven Orchestrated, Concerted, Comprehensible shown on the Wheel.

organization. However, innovativeness in professionals is very essential to motivate others for adoption of modern technologies. (2) Decision making ability: Considered as the nature of decision making either individually or consulting with others while performing activities. It is the degree to which an individual justifies his selection of most efficient means from among the available alternatives on the basis of scientific criteria for achieving maximum profits. Hence, decision making ability is very important among professionals to motivate for adoption of modern technologies. (3) Achievement motivation: Every man has a desire to achieve certain things in their life. Achievement motivation is considered as the extent to which an individual is oriented towards maximizing profits. Achievement motivation as a social value that emphasizes a desire for excellence in order for an individual to attain a sense of personal accomplishment. So achievement motivation increases efficiency of professionals in use of modern technologies.(4) Information seeking ability: It refers to the frequency of contact by professional with various information sources. This is the pattern by which an individual gets information either on his/her own seeking or as a consequence of his being a part of the network. This component is important for use of modern technologies by professionals. (5) Risk taking ability: Some take more risk, some others take moderate risk and many hesitate to take risk. Risk taking ability considered an individual orientation towards risk and uncertainity in adopting new ideas and courage to face the problems. Use of modern technologies demands ability to take risk by professionals. (6) Coordinating ability: In order to complete the required work in stipulated period, one has to harmonize and synchronize the various activities for better profit. It is an individual co-ordinates action in a time dimension. This ability helps to increase the efficiency of professional in motivating farmers for adoption of modern technologies. (7) Leadership ability: To get things done properly, a professional has to initiate the action, motivate the followers and decision should be taken. It is an individual initiates or motivates the action of the other fellows. Hence, leadership ability is an important component in professional to motivate others for adoption of modern technologies.

5. Skill Development among Professionals for Application of Information System to Increase Sustainable Agricultural Production

Under the current agrarian issues to meet the requirement of emerging problems of sustainable agriculture development the professional concerned should be trained properly in specific skills required for conducting an effective training of farmers. The skill as ability to do things, to effectively apply knowledge and personal aptitudes and attitudes in work situation. However, the concept of skill concerns the ability to use ones knowledge effectively and rapidly in execution of performance. More generally it is an acquired power of doing some thing competently. Skills, notably professional skills are becoming increasingly important. Further, sustainable Agricultural Information System to day calls for professional skills in its application. Further, seven skills are identified for effectiveness among professional. However, Narayana's Wheel Model (Figure 7.4) describes seven orchestrated, concerted comprehensible skills required among professional for efficiency like (1) Technical skill is the ability of the professional to use any technique or method or equipment or

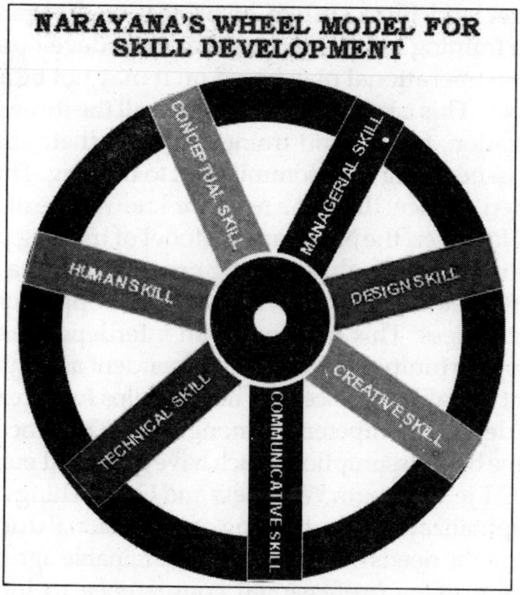

Figure 7.4: Seven Orchestrated, Concerted, Comprehensible shown on the Wheel.

product or process as a tool in the context. (2) Human Skill is the ability of professional in motivating other people involved in with thorough understanding while working with them as a team. (3) Conceptual Skill is the ability of professional coordinating and integrating of all the activities with visionary outlook. (4) Managerial Skill is the ability of professional in planning, organizing, directing, leading, reporting and budgeting and reviewing the work of other people involved. (5) Design Skill is the ability of professional in finding out a workable solution to problems. It requires deliberate efforts to develop solution. (6) Creative Skill is the ability of professional in generating new ideas or in doing things already done in a new way. (7) Communicative Skill is the ability of the professional to disseminate and make others to adopt technologies at different levels using series of methods over a period of time.

6. Training of Professionals for Efficient Use of information System to Increase Sustainable Agricultural Production

Training is a planned and systematic effort to increase professional competency. Further, to enable the professional to increase knowledge, to improve skills, to inculcate appropriate attitude and develop appropriate attributes to serve better. Several training models are used by the organizations to influence the professional to make desirable changes in their behaviour to achieve the objectives of the organization. Further, observed that training is a building process, to reflect this, a good course is organized in ascending order of complexity. However, understanding of modern technology and deliver it to users in a usable form along with monitoring of activities needed to implement and evaluate its usefulness are urgently needed. The information has to be integrated with available communication methods to suit the resource positions of institutions and time following the integration. Further, good linkage have to be established with inter and intra system of professional organizations. Further, experience gained in training so far indicates that mere development of conceptual understanding and an operational plan based on it may not be adequate. Training has to be made to work. This can happen only when all the three parties involved in training like organization, trainer and trainee – join in their effort and make it to work. An important issue facing us is commitment to training. This is required and it is seldom well realized. Hence, there is a need for knowledge of training models to train professionals. However, the Narayana's Model of training process (Figure 7.5) for training professional explains that training process may be a temporary system but the trainer and trainee both learn through various opportunities available for checking their effectiveness. This training as an interdependent and interrelated process. Here lot of opportunity is given for independent and intervening variables to become dependent variables. Hence, this model helps to increase the efficiency of professionals and to develop competency among them. The process of training must start by questioning the basic assumption which have governed our training approach. So an analysis of SWOT ie., Strength Weakness and Other Things called for to enable as to have new conceptualization. The training of professional during the millennium must take in to account the needs of broad based sustainable agriculture information system to introduce greater professional competence to increase sustainable agricultural production.

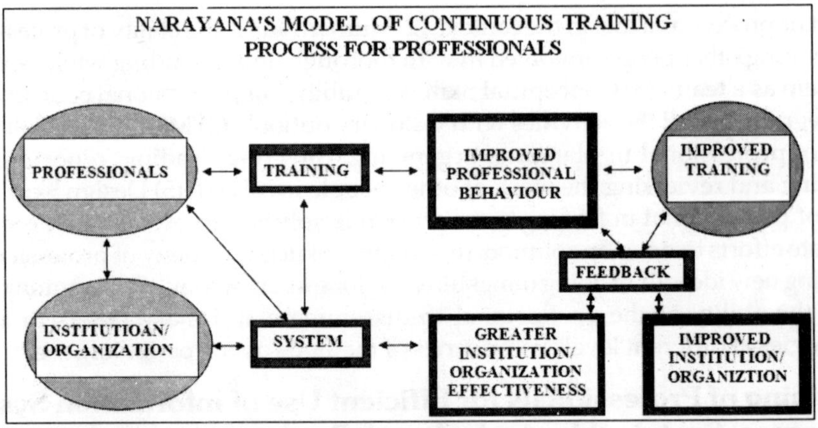

Figure 7.5

7. Human Resource Development to Use information System Efficiently to Increase Sustainable Agricultural Production

During the last five decades of development, growth of developing countries is directly related to their human resource bases. The countries which have given good performance are the countries which have made significant investments in Human Resource Development (HRD). There is an over whelming evidence that human capital is one of the key factors for adoption of modern technologies in Developed countries. Further, HRD is widely regarded as the single most important resource for faster adoption being attempted in the developing countries. Hence, there is need for training to develop human resource at various levels in organizations for efficient use of modern technologies. Human resources are assuming increasing significance under the current agrarian issues. However, experience in the past has indicated that HRD among professional is lacking in our country as revealed by large number of research and evaluation studies. How, it should be done is explained in Narayana's Algebraic Model of HRD (Figure 7.6) among professionals. HRD = HRS + HRT + HRU *i.e.*, HRD = HR (S+T+U) *i.e.*, Selection of Human Resource, Training of Human Resource and using of Human Resource profitably are urgently needed for adoption of emerging technologies to increase sustainable agricultural production. This is a challenge that needs to be tackled immediately.

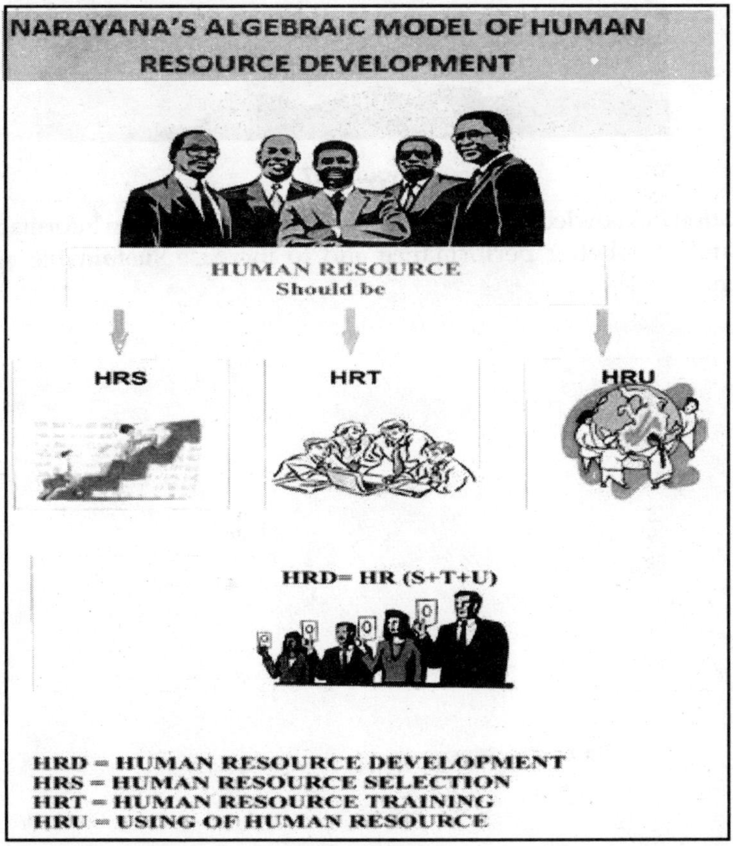

Figure 7.6

8. Strategy for Better Performance to Increase Sustainable Agricultural Production

Narayana's Algebraic Formula explains how to get better performance from the professional. Which indicates that P=A+W *i.e.* P= performance, A= ability, W= willingness, Ability is further given as A=K+S *i.e.* K= knowledge, S= skill and further W=V+A *i.e.* V= values, A= attitudes. For better performance all the above *i.e.* knowledge, skill, values, attitudes are equally important. However, wisdom should be developed

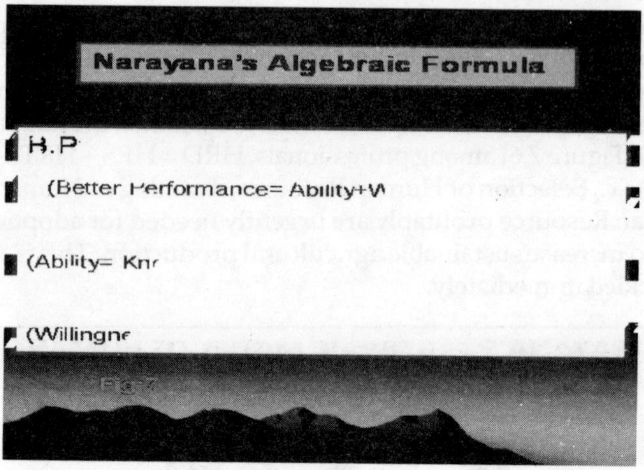

Figure 7.7

for application of knowledge and skill in sustainable Agriculture Information System appropriately for better performance and to increase sustainable agricultural production.

Chapter 8

Evolving an Inclusive Extension System for Enhancing Rural Household Income

R.P. Singh 'Ratan', A.K. Singh and Niva Bara

Introduction

Agriculture plays an important socio-economic role by providing livelihood opportunities to a vast majority of population. In disadvantage areas it is key for ensuring food and nutritional security, economic development and generation of employment opportunities. During the first decade of 21st Century there emerged two contrasting trends in India that the country has been increasingly recognized as a global power in key economic sectors with a consistent high economic growth and the performance of agriculture sector has been found to be dismal causing concerns for future food and nutritional security as well as household income security. This trend is very much evident in rainfed areas where agriculture is practiced done by the weaker sections operating the small farm production systems, who have also not been benefitted by the key economic reforms. Growing competition for land, water and energy is affecting our ability to produce food. Therefore, conserving natural resources, maintaining biodiversity and accelerating agricultural growth are considered to be of paramount importance in the present and future contexts to ensure sustainable and equitable food, nutrition and household income security. In this scenario present day agriculture has to go beyond mere food production to be a propellant of economic development by being market-oriented. Thus agricultural extension has to play a crucial role in ushering in such a transformation in agriculture to meet the future challenges.

The recent reforms in Indian agricultural extension systems like integrated approach, participatory planning and management, group mobilization, etc. call for major change in institutional setting (Sulaiman and Hall, 2003). These changes should eventually lead to "learning rather than command and control" as the philosophy. However, these reforms have not produced the desired effects, implying thereby that instead of searching for outside solutions to typical micro-level problems in rural and agrarian India, the public extension must learn to look for such solutions inside the client systems and if required replicate the same to achieve the economics of scale. This requires that clients be viewed as active partners not just passive recipients of extension.

Emerging Innovations in Extension

To meet farmers changing need for information and advisory support, extension should engage with a wide range of issues related to agriculture (Van den Ban, 2000). These include market, credit and insurance, in addition to technology and research services and making arrangements for the supply of inputs. The field of extension now needs to address a wide range of activities *viz.* linking farmers to markets (Neuchatel Group, 2002), reducing vulnerability and empowering the rural poor (Farington *et al*, 2002), developing micro-enterprises (Rivera *et al*, 2001), poverty reduction and environmental conservation (Alex *et al*, 2002) and strengthening and supporting farmer organizations (Sulaiman and Hall, 2003). Over the last decade, the need for extension functionaries to reorient their skills and activities towards community mobilization, conflict management, problem solving, education and human development has been emphasized (Farrington *et al*, 1998, Sulaiman and Van den Ban, 2000).

The goal of extension has, therefore, is moving from technology dissemination to system management, building coalitions of different stakeholders, leading the innovation agenda and building new learning organizations. Public sector extension needs to make conscious efforts to learn from ongoing institutional experiments and should be restructured with the necessary skills and capacities to integrate information and expertise available in different organizations (Sulaiman and Van den Ban, 2003). Emerging extension innovations tried over space and time mainly fall under four broad models, namely, decentralization (shifting responsibility of extension to local governments), privatization (shifting substantial or total ownership and control to the private sector), cost recovery and cost sharing, participatory and demand-driven extension (group approaches, farmer field schools and farmer consultations/participation during implementation).

The basic approach to agricultural extension in 11th Five Year Plan involved focus on extension reach to small, marginal farmers, farm women and disadvantaged groups; strengthening and upscaling new institutional mechanism with adequate funds for extension activities, dedicated manpower at different levels and infrastructure support; focusing on demand-driven and farmer-oriented training strategies; ensuring women participation in all the trainings programmes at all levels and organizing specialized training for women trainers; introducing gender budgeting in agricultural extension programmes and sensitizing all the stakeholders in gender issues; farmer-led-extension as an instrument for participatory research and knowledge management leading to build a cadre of grass-root farmer trainers, and to facilitate commodity interest groups; agri-preneurs identification and training under KVKs or agri-clinics and agri-business centres recognizing the cutstanding farmers through awards; provision of adequate quality extension manpower including the private service providers (agripreneurs, input dealers, NGOs, programme producers of print and mass media, etc.) to address agricultural diversification, intensification, commercialization and sustainability issues, understanding information needs of different clients (farmers, farm women, labourers, youth, etc.) through continuous capacity building to also have skills of group formation, negotiation, conflict resolution, social mobilization, management of Common Property Resources (CPRs), use of IT, documentation of success stories and replication to improve the effectiveness of the delivery system; Integrating Research-Extension-Farmer-Market linkages to be addressed by undertaking research and extension activities through the participatory technology development mode, creating a Research-Extension-Farmer and Market coordination at State level to take necessary policy initiatives to enable and establish linkages; promoting farmer organizations around the commodity and federating them at block/district/state level and linking the economic activities and markets for collective bargaining to sustain the developmental efforts; following farming system approach explicitly by farmers' needs through an understanding of the existing farming systems to determine the research and extension agenda and, convergence of extension services between public and private sector extension service providers on Public-Private-Partnership (PPP) mode at the district, block and village levels to improve the performance of various stakeholders.

Mass Media and ICT for Inclusive Extension

Considering the diversity of agro-ecological systems and more than 120 million farmers involved, the extension system is faced with a great diversity of needs, opportunities and prospects. Therefore, to respond successfully to the new challenges posed, greater attention will have to be paid to information-based technologies and strengthen means of dissemination to transmit the information to farm families. Effective use of mass media and ICT could be one of the possible means for bridging knowledge deficiency among the farming community at a faster rate. The use of Information Technology in extension enables the extension workers to be more effective in meeting the information needs of farmers and farm women. A dedicated TV channel on agriculture, usage of strengthened network of community radio and FM radio to focus on location-specific problems of farming community and technology transfer and strengthened Call Centres in agriculture such as Kisan Call Centes provide for problem solving in agriculture through toll-free numbers are recommended. For speedy transformation of technology based societies, information needs to be packed in digital video-based web environments for dissemination through the Village Knowledge Centres / Information Kiosks and Digital Media Forums as well as through CDs and multimedia presentations. Information Communication Technologies (ICTs) aimed at utilization of need-based real time information for decision making and empowerment leads to development.

Guiding Principles for Inclusive Extension System

The following guiding principles suggested by Swanson (2008) serve as tenets for evolving inclusive extension in India

- ✰ Economic efficiency;
- ✰ A careful match between the comparative advantages of organizations and the functions they perform;
- ✰ Enabling policy and governance environment;
- ✰ Cear spread of costs;
- ✰ Careful assessment and optimal mixing of funding and delivery mechanisms;
- ✰ Pluralistic and participatory approaches;
- ✰ Effective linkages among farmers, educators, researchers, extensionists and other stakeholders;
- ✰ Building human and social resources; and
- ✰ Sound monitoring and evaluation to project better impacts.

Key Features of Inclusive Extension System

In order to meet the next generation challenges and to harness the vast opportunities, future extension systems would be evolved around the following key features:

- ✰ Responsive to the ever-changing client needs (informational, technological, capacity building/educational, social mobilization, etc.)
- ✰ Pluralism or multi-agency approach as the operating philosophy
- ✰ Technology-enabled extension methodologies (modern IT and ICT-led methods would be increasingly used in addition to conventional methods)
- ✰ Partnerships of various types (public-public, public-private, private-public and private-private) to harness the comparative advantages of different extension service providers
- ✰ Micro-strategies aimed at promoting grassroot level linkages by innovative strategies (best-fit rather than best practices approach)/innovations rather than nation-wide prescriptive strategies
- ✰ Concentrating more on what extension can do and do best - education, facilitative role more than technology transfer and input supply rather than embracing everything decided by others.

Suggestive Model of Next Generation Agricultural Extension in India

In view of changing partnerships and proprietary claims, there is a need to bring system approach in extension and transfer of technology and to look for an extension model that would involve various extension agencies like KVK, Agricultural

Technology Management Agency (ATMA), line departments and private agencies, etc. in the district. This framework is based on the following premises:

- Core operating philosophy of next generation agricultural extension system will be guided by its classical manifestation, education, facilitation and empowerment (helping clients to help themselves).
- District will be the core planning and operating unit.
- Systems approach to link all extension service providers (ESP) at district level and below with innovation systems (formal and informal research systems), client systems and other support systems (inputs, credit, power, road, telecommunications, etc.).
- KVKs will be the main agencies to connect all the ESPs with other systems directly by ways of functional linkages or by facilitating regular interfaces, initial hand-holding in forging partnerships and also by referral/regulatory role for diagnostic and prescriptive services.
- Within the purview of the existing mandate, the KVKs will function as nodal extension agency. However, the KVKs will have flexibility to realign its mandated activities in line with the development niches of the district, and also its own strengths.

The model proposes KVKs as the basic platform at district level to spearhead extension on the premises suggested above. The backend support for KVKs is visualized in the form of National Agricultural Innovations and Knowledge Gateway. KVKs would customize the innovations and knowledge repository of this gateway to help the clients on a group approach (FIGs, CIGs, FOs, SHGs, etc.). KVKs would facilitate and help these groups to connect with the other ESPs. KVKs would use Knowledge Infomediaries (KnIMs), to be identified jointly by Panchayati Raj Institutions and KVKs as master trainers and KVK ambassadors to multiply and magnify the capacity building and other extension activities using Farmer School concept. KVKs would help connect National Agricultural Innovation and Knowledge Gateway to the future generations by forging active partnerships with schools in the district.

The possible strategic options for this model are

- Decentralization/devolution/democratization
- Top-down/bottom-up/convergence approach for knowledge and innovation dissemination
- Free of cost/cost sharing/full cost recovery as fiscal considerations
- Segmentation (area/client type/commodity), targeting (customizing knowledge and innovations and costing options) and positioning (differential pricing, communication, media-mix, etc.) of ESPs.

Critical Success Factors for Next Generation Extension System in India

The following factors would be the corner stone of the next generation extension system in India.

- *Pluralism:* Acknowledge and integrate pluralism in extension (harmony and synergy among various ESPs).
- *Professionalism:* Move away from bureaucratic to more professionai operating philosophies and systems. This include core values of the systems and people constituting these systems, professional identity, professional HRD and capacity building.
- *Pragmatism:* Ability to 'think global and act local' to consider and cater to real needs of the clients and not the greed of clients/donors.
- *Partnership:* Shift from competitor approach to strategic issue-based alliances, partnership and consortium approach.
- *Prudence:* External demands and internal responses would call for judicious and efficient use of resources to deliver the best results (outputs and outcomes).
- *Pride:* A sense of commitment, devotion, dedication, and discipline to promote pride in serving those (millions of small, marginal and landless farmers and other downtrodden sections of rural India) who strive to ensure the food security of the nation.

Other critical factors that influence the success of this market-driven extension approach include the followings:

- Helping farmers, including farm women, who have similar resources and interests, organize into producer groups within each community.
- Having access to reliable markets and market information about which crops and products have sufficient economic potential to be produced and marketed in the different agro-ecological zones within each district.
- Having access to production inputs (*e.g.*, seeds) and training in the production and other or products and to meet market specifications. It should be noted that urban consumers are increasingly influenced by global food preferences. Therefore, it may be necessary to secure new varieties form international sources (*e.g.*, private-sector companies), because these planting materials may not be readily available locally.
- Product identification and certification process (knowledge and access).
- It is important to remember that a market-driven extension approach helps farmers move incrementally toward agricultural diversification. Small-scale subsistence farm households do not stop producing the basic food crops needed for home consumption. Rather, they allocate a small amount of their land to produce a specific high-value crop (*e.g.*, fruits or vegetables) or products of backyard poultry, honey, mushrooms, etc. and after they work out the necessary production and marketing practices. Then, they begin scaling up the production of these crops or products, based largely on profitability.

To Conclude

The changing farmers' priorities form production to income has shifted the extension emphasis to help farmers to organize themselves and to link them markets. Other important roles that extension needs to play are (i) building coalitions of different stakeholders; (ii) promoting platforms for information sharing; (iii) experimenting with and learning new approaches; and (iv) acting as a 'bridging organization' that provides access to knowledge, skills and services from a wide range of organizations including research institutions.

Bringing farmers, processors, retailers and other stakeholder together to support modern agricultural practices and convergence of extension efforts especially at the district level and below would be required. Common Service Centres (CSCs) of the Department of Information Technology, Government of India and those set up by the state governments and private initiative programmes will be evolved for inclusive and broad-based development.

Empowering farmers with the right information at the right time and place is essential for improving the efficiency and viability of small and marginal holdings. Mass media, particularly the radio, television and local language newspapers will be used to play an important role in this regard.

References

Alex, G., Zijp, W. and Byerelee, D. (2002). Rural Extension and Advisory Services – New Directions. Rural Development Strategy Background Paper No. 9, Agriculture and Rural Development Department. Washington, DC: The World Bank.

Farrington, J., Christopher, I., Kidd, A.D. and Beckman, M. (2002). Extension, poverty and vulnerability: the scope for policy reform, Working Paper, 155. UK:ODI.

Neuchatel Group (2002). Common framework on financing agricultural and rural extension, Neuchatel Group. Linadu: Swiss Centre for Agricultural Extension and Rural Development.

Rivera, W.M., Crowder, L.V. and Qamar, K. (2001). Agricultural and rural extension worldwide: options for institutional reform in the developing countries. Extension, Education and Communication Service. FAO. Rome.

Sulaiman, R. and Hall, A. (2003). India: The emergence of extension – plus: future for extension beyond technology transfer? In: W.M. Rivera and G.E. Alex (eds.), Extension Reform for Rural Development, Washington, DC: The World Bank.

Van den Ban, A. (2000). Different ways of financing agricultural extension. AgREN Network Paper No. 106. Overseas Development Institute. London: ODI.

Chapter 9
Ground Realities of Technology Transfer: Points for Intervention

R.P. Singh

Economic survey 2011, talks of robust growth and steady physical consolidation as the Hallmark of the Indian economy. After all, with a growth of 5.4 per cent for agriculture and the allied sector on the back of the increase in food grain production this year, countries GDP has been worked out at 8.6 per cent. Agro technology transfer among the states and regions in the country has constraints and problems additional to those associated with the diffusion of innovations (L. D. Swindale, 1978.). Site factor constraints the appropriateness of the technology, social, economic and institutional constraints must all be recognized and conceptualized. Environmentally Sound Technologies (EST) has the fewest constraints and has been a major means for agro technology transfer.

The technology is complex and complete involving many factors and many new practices, however, as scientists have shown, it can be adopted piecemeal although, none of the pieces bring the advantage that the entire technology does. How, now, do we transfer this agricultural technology?

We need a revolution. The present paper will discuss the successful transfer and uptake of Environmentally Sound Technologies. It is time to reverse the clock and provide a ray of hope.

Technology Transfer in Support of Sustainable Development

Technology transfer must be recognized as a broad and complex process if it is to avoid creating and maintaining the dependency of the recipient, and if it is to contribute to sustained and equitable development. The end result for the recipient must be the ability to use, replicate, improve and, possibly, re-sell the technology. Transfer of technology is more than just the moving of high-tech equipment from the developed to the developing world, or within the developing world. Moreover, it encompasses far than equipment and other so-called "hard" technologies, for it also includes total systems and their component parts, including know-how, goods and services, equipment, and organizational and managerial procedures. Thus technology transfer is the suite of processes encompassing all dimensions of the origins, flows and uptake of know-hw, experience and equipment amongst, across and within countries, stakeholder organizations and institutions.

If the transfer of inadequate, unsustainable, or unsafe technologies is to be avoided, technology recipients should be able to identify and select technologies that are appropriate to their actual needs, circumstances and capacities. Therefore, a key element of this wider view of technology transfer is choice. There is no single strategy for successful transfer that is appropriate to all situations. Desirably a technology recipient will choose a technology which at least meets the definition of being "environmentally sound". Suitable technologies are technologies that have the potential for significantly improved environmental performance, relative to other technologies. ESTs protect the environment, are less polluting, use resources in a sustainable manner, recycle more of their wastes and products, and handle all residual wastes in a more environmentally acceptable way that the technologies for which they are substitutes. Preferably a technology recipient will go even further, and select a "sustainable technology" – *i.e.* a technology that is not only environmentally sound but also economically viable and socially acceptable. Such technologies contribute to the three pillars of sustainable development.

Elements for the Transfer and Uptake of Environmentally Sound Technologies

Environment/Situation

Technology transfer does not take place in a vacuum. The performance of a given technology is dependent on a wide range of factors, making identification of an environmentally sound or otherwise appropriate technology somewhat problematic. For example, a technology that is assessed to be environmentally sound in a given locale, culture, economic setting or stage in its life cycle may not be in another. Its performance may be influenced markedly by the availability of supporting infrastructure and by access to the expertise necessary for its management, maintenance and monitoring. Moreover, a technology that qualifies as being environmentally sound at one point of time, may not do so at another – the performance criteria against which it is assessed may change as a consequence of new information or changing values or attitudes; a technical breakthrough may give rise to more desirable alternatives. It is therefore vital that recipients and users of a technology are able to choose an option that meets their specific needs and capacities, while also being environmentally sound in its operating locale and over its operational life cycle. It is, of course, highly desirable that the technology is also found to be economically viable and socially acceptable, and hence sustainable.

Obstacles to Remove

There are many barriers to successful technology transfer. All along the transfer path, from the supply side of technology transfer (the innovators and developers) to the demand side (the recipients and users), impediments occur at every node and, due to restrictions on the movement of information and materials, for every linkage in the technology transfer chain. While some generalizations are possible, the specific nature and severity of the challenges depend on the prevailing circumstances, varying with the type of technology, its specific application and the characteristics of the technology providers and recipients. Examples of challenges include shortfalls in technology creation and innovation, underperformance in technology sourcing, sub-optimal enabling environments, and insufficient and unverified information. Small and medium enterprises are disproportionately impacted by these challenges.

Choice

A key aim of barrier removal, that is of facilitating technology transfer, is ensuring that technology recipients and users are able to make informed choices by being able to identify and procure the most appropriate (in environmental and preferably also in economic and social terms) technology for a given application in a given locale. Several requirements must be met, including:

- ☆ Needs well defined, documented and understood;
- ☆ Several technology alternatives, all of which are well and reliably characterized in terms of environmental and economic performance and potential social impact;
- ☆ Rational and functional methods (decision support tools) that facilitate choice of an optimal technology; and
- ☆ Capability to make the chosen technology fully operational, so that it fulfils its potential, and meets the identified needs, without detrimental side effects, including during decommissioning.

Certainty

A lack of certainty, and the consequential high levels of risk, both real and perceived, are recognized as major impediments to the successful establishment and ongoing operation of functional markets for ESTs. The common perception that many ESTs are "emerging", and hence "unproven", means there is little confidence in their economic, commercial or technical viability. Removing barriers to technology transfer often translates into increased certainty, and decreased risks, for the key stakeholders such as the developers, suppliers, financiers, insurers, recipients and regulators. One example is ensuring access to sufficient, verified information. Risk assessment and management capabilities for financial institutions are also of special importance.

Policy instruments can also be used to enhance certainty, in two principal ways:

- ☆ Through consistency in policy goals and measures over time, and with long lead periods for substantive changes in policy directions and the measure that implement them; and
- ☆ Use of policy instruments to reduce regulatory, investment and other uncertainties in the market.

Macroeconomic conditions that favour technology transfer include those which will deliver low inflation, stable and realistic exchange and interest rates, pricing that reflects the true (marginal and fully internalized) costs of material, energy, labour and other inputs, deregulation, free movement of capital, operation of competitive markets, open trade policies and transparent foreign investment policies.

Effective Communication

The technology transfer chain is often long, in terms of both distance and time. Effective communication is thus another essential ingredient in the recipe for successful technology transfer. Efficient and effective two-way communication and cooperation between key stakeholders will do much to remove barriers. Information management systems, knowledge management tools and formal and informal networks, both centralized and dispersed, can all make important contributions. Technology transfer often involves a dissonant mix of informal actors (*e.g.* innovators) working in formal and highly regulated settings. Effective communication is a requisite to harmonizing the contributions to the processes of technology transfer being made by diverse players.

Creating Ability

Enhancing the transfer of technologies that support sustainable development is largely about creating favourable circumstances for technology transfer – ensuring all stakeholders have the ability (potential and realized) to fulfil their roles and meet their responsibilities, expeditiously. Generally speaking, government is the principal player in creating an enabling environment for technology transfer, but financial and insurance institutions and international organizations can also be influential.

Circumstances which are supportive of technology transfer include:

- ✰ Open and competitive market;
- ✰ Comprehensive and credible specifications on the technology performance;
- ✰ Financiers who are at least technology neutral;
- ✰ The most cost competitive technology also has the most favourable environmental and social performance specifications; and
- ✰ Policy risks are addressed.

All key players and stakeholders must have the necessary knowledge and skills to perform the roles and tasks expected of them. High levels of awareness, motivation and empowerment within the public and private sectors and in civil society will help ensure that people, communities and wider societies are able to adapt continuously to new circumstances and challenges that drive and arise from technology transfer.

Effective and efficient national and regional systems of innovation, research ad development should be in place, to facilitate such procedures as adaptation of traditional technologies for use in current settings. The enabling environment also benefits from policy implementation that fosters an appropriate mix of government and private sector investment in ESTs and address such issues as lack of access to appropriate sources of capital, high or uncertain inflation or interest rates, subsidized or average-cost (rather than marginal-cost) prices for material and energy inputs, high import duties, uncertain stability of tax and tariff policies; investment risk (real and perceived), loss of rights to intellectual property and to productive resources and risk of expropriation.

Commitment

If there is to be an improvement on the last decade or so, where technology transfer failed to deliver the anticipated and much needed advances in development and sustainability, we must make a commitment to overcoming the challenges, providing technology users with the choice they deserve and desire, increase certainty, thereby reducing risks, enhancing communication between technology transfer stakeholders and building and strengthening the enabling environment and thus the capacity for technology transfer.

Key actions that will foster technology transfer include:

- ☆ Needs assessments, including identification of shortcomings in the enabling environment, with relevant organizations and agencies helping to address these;
- ☆ Evaluation and strengthening of policies that influence the enabling environment;
- ☆ Greater communication and interaction between key parts of government
- ☆ Intra- and inter-governmental coordination, cooperation and assistance;
- ☆ Protection of intellectual property rights and legal contracts;
- ☆ Political support for programmes and institutions that foster technology transfer;
- ☆ Seed investment programmes to stimulate private sector investment;
- ☆ Capacity enhancement for major stakeholders;
- ☆ Delineation of the roles of the private and public sectors in both developed and developing countries;
- ☆ Economic incentives targeting industries that have the potential to make critical and major contributions to technology transfer; and
- ☆ Ensuring that technology transfer initiatives are compatible with national sustainable development agendas;
- ☆ Increase communication among technology transfer bodies across various multi-lateral environmental agreements (MEAs) with a view to leveraging limited financial and human resources on issues of common interest, integrating and strengthening regional and country level activities through information sharing and joint activities and providing a platform for multilateral approaches and consistency in technology transfer.

As part of its catalytic and facilitation role in creating and implementing strategies for transformation and change, UNEP and its partners are working together to develop and implement a strategic framework for promoting the adoption and use of ESTs. UNEP is well-positioned to provide an effective platform for meaningful interaction and dialogue in support of the harmonization of assessment approaches and methodologies related to ESTs. To demonstrate the benefits of ESTs, UNEP has established an EST Initiative with a number of partner organizations. A key objective is the transparent and credible acquisition and reporting of environmental performance information related to technologies.

References

SWINDALE, L. D. 1978. A soil research network through tropical soil families. Pages 210-218 *in* Soil re-source data for agricultural development, ed. L. D. Swindale. Honolulu: Hawaii Agricultural Experiment Station.

Chapter 10

Impact of Watershed Development Programmes on Crop Productivity

Biswajit Mondal, Alka Singh and I. Sekar

Introduction

India accords high priority to watershed based interventions as a strategy for improving livelihoods and sustainability in drought-prone areas. This goes beyond merely conservation activities in terms of construction of structures and emphasizes importance of human dimensions, strengthening local institutions, income-generating technologies and markets to improve livelihoods (Kerr *et al.*, 2000 and Reddy *et al.*, 2004). Most watershed projects are being implemented with the twin objectives of natural resource conservation and enhancing the livelihoods of the rural poor through enhancement of production levels (Sharma and Scott, 2005).

Various evaluation studies carried out at different watershed located in different agro-climatic conditions have shown that watershed programmes help to increase agricultural productivity besides conserving natural resources. While there are no disputes about the beneficial impact of watershed projects, one important issue that has not been answered by the researchers that how much portion of this benefit (productivity in question here) is purely due to watershed technology. If watershed technology is an improvement over control areas, then its' effects in terms of gain in productivity should have occurred in two stages. Initially, more output is made available from the existing resource base under the new production technology (in this case watershed technology). This is the efficiency component, reflected in the shift in the production function. Second, an adjustment component of technological change is evident in the movement along the new production function. This movement along the new production function follows from the efforts of the firms to adjust to disequilibrium caused by the new level of efficiency. It is worthwhile, therefore, to decompose the total difference in output into its' causative factors of the differences in the levels of input use and technological efficiency.

Decomposition method was first used in Indian Agriculture by Bisaliah (1977) to evaluate the effect of technical change due to introduction of Mexican wheat in Punjab farms. Subsequently, the method was used by several other workers as Umesh and Bisaliah (1986), Lalwani (1989), Kiresur *et al*. (1995) and Kiresur and Ichangi (2011) to evaluate the technological gap in crop as well as dairy farming. There is very scanty literature on the use of decomposition analysis to break up the output growth of watershed technology into various causative factors. Hence, it was attempted to ascertain the impact of watershed development programmes on crop productivity and dissociate the changes purely into watershed technology.

Methodology

Bundelkhand region of Madhya Pradesh state was selected for the present study. Located in a hot and semi-humid region between the Yamuna and Narmada, Bundelkhand region is backward relative to other regions of the state (Inter-Ministerial Central Team Report, 2008). Eight watersheds implemented under different types of government departments (GO) as well as non-governmental organizations (NGO) as project implementing agencies (PIAs) were selected from the region. To make a comparative study, one control village from the contiguous area of each selected watershed where no watershed development activities were carried out, was also chosen. A total of 240 sample households were chosen from the selected villages and survey was conducted during 2010-11 for detailed investigation. The primary data pertaining to the socio-economic characteristics of respondents along with the crop cultivation details were collected by personal interview of the respondents with the help of pre-tested comprehensive schedule particularly designed for the study. In absence of the benchmark data for watershed villages, input-output data from control villages were used as a proxy for the 'old technology' in the decomposition analysis.Considering the differences in crops and their varieties among the farmers, gross returns and input usages per unit area in monetary terms were used for functional analysis. For accomplishing the task of decomposition analysis, Cobb-Douglas production function has been used as follows:

$$Y_i = A_i N_i^{ai} S_i^{bi} F_i^{ci} M_i^{di} K_i^{ei} \varepsilon^{ui} \quad (1)$$

Where,

Yi = Gross returns from crop enterprises, in Rs./ha

N_i = Human labour input, in Rs./ha

S_i = Value of seed, in Rs./ha

F_i = Value of fertilizer and manures, in Rs./ha

M_i = Machine labour input, in Rs./ha

K_i = Miscellaneous capital expenses, in Rs./ha

A_i = Constant term of scale parameter

u_i = Error term with zero mean and finite variance

a_i, b_i, c_i, d_i, e_i = Partial output elasticities of human labour, seed, fertilizer, machine labour and capital expenses, respectively.

Following the above model of production function the same for control areas and watershed areas were specified as below:

$$Y_1=A_1N_1^{a1}S_1^{b1}F_1^{c1}M_1^{d1}K_1^{e1}\varepsilon^{u1} \quad (2)$$

$$Y_2=A_2N_2^{a2}S_2^{b2}F_2^{c2}M_2^{d2}K_2^{e2}\varepsilon^{u2} \quad (3)$$

Definitions of variables and parameters in (2) and (3) are the same as in (1).

The decomposition equation following *Bisaliah (1977)* from the above production function was specified as below: $\text{Log}\,[Y_2/Y_1] = \log\,[A_2/A_1] + [(a_2\text{-}a_1)\log N_1 + (b_2\text{-}b_1)\log S_1 + (c_2\text{-}c_1)\log F_1 + (d_2\text{-}d_1)\log K_1] + [a_2\log(N_2/N_1) + b_2\log(S_2/S_1) + c_2\log(F_2/F_1) + d_2\log(K_2/K_1)] + (u_2-u_1)$ ....(5)

The decomposition equation (4) involves decomposing the natural logarithm of the ratio of output at watershed to the output at control areas. It is approximately a measure of percentage change in output with the introduction of watershed development programmes. The first bracketed expression on the right hand side is a measure of percentage change in output due to shift in scale parameter (A) of the production function; is attributable to the neutral component of technology. The second bracketed expression, the sum of the arithmetic changes in output elasticities each weighted by the logarithm of that input used, is a measure of change in output due to shifts in slope parameters (output elasticities) of the production function (non-neutral component of technology). The third bracketed term refers to the gap attributable to differences in the input use weighed by the slope coefficients of the productivity function fitted for watershed areas.

Results and Discussion

Changes in productivity in the sampled watersheds due to watershed based interventions Different soil and water conservation activities were undertaken in the sampled watersheds as per the needs and priorities of the watershed community and their technical feasibility. Constructions of different types of gully control structures were made and runoff control measures like vegetative hedges were developed to arrest erosion/stabilize gullies. Creation of water resources potentials was undertaken through construction of water harvesting structures of different sizes and capacities, renovation/rejuvenation of existing structures and construction of new wells. Other types of structures like percolation tanks, well recharge pit, sunken ponds, etc, were also constructed. Horticultural plantation, afforestation and fodder development activities were also adopted on farmer's fields/boundaries as well as on non-cultivated land for income generation, wasteland reclamation and fodder production. The cumulative effect of all the above-mentioned land-based interventions was reflected through favourable changes in various bio-physical indicators/indices like irrigation status, cropping pattern and intensity which ultimately led to increased productivity of almost all the crops grown over the pre-project years (Table 10.1).

Table 10.1: Changes in Productivity in the Sampled Watersheds due to Watershed Based Interventions

Crops	*Pre-Project Yield (kg/ha)*	*Post-Project Yield (kg/ha)*	*Per cent Change*
Rabi crops			
Wheat	885	1447	63.50
Gram	720	932	29.44
Lentil	411	524	27.49
Mustard	448	631	40.85
Linseed	402	476	18.41
Kharif crops			
Soybean	652	918	40.80
Urad	342	395	15.50
Paddy	714	919	28.71
Arhar	499	627	25.65
Sesame	213	353	65.73

Estimates of Production Function (per ha)

All the estimated production functions were significant as evidenced by the significance of F-values at 1 per cent level (Table 10.2). A perusal of the production

Table 10.2: Estimates of Production Function (per ha)

Variables/Particulars	*Regression Coefficients*		
	Watershed Area	*Control Area*	*Pooled*
Human labour (N)	0.16 (0.19)	0.07 (0.25)	0.28 (0.18)
Seed (S)	0.15 (0.13)	0.01 (0.20)	–0.07 (0.13)
Fertilizer (F)	0.00 (0.14)	0.14* (0.20)	–0.11 (0.14)
Machine labour (M)	0.32*** (0.11)	0.39** (0.16)	0.40*** (0.11)
Capital (K)	0.25*** (0.08)	0.25** (0.11)	0.32*** (0.08)
Intercept	3.31* (1.89)	3.48 (2.67)	3.57* (1.86)
R^2	0.31	0.27	0.23
F-value	9.02***	7.58***	12.94***
Sample size (n)	108	109	217

Figures in parentheses are the standard errors.

***, ** and * indicates the coefficient are significant at 10, 5 and 1 per cent probability levels.

function estimates reveals that the coefficients of all the inputs *viz.*, human labour (N), seed (S), fertilizer (F), machine labour (M) and capital expenses (K) were invariably positive for both watershed and control areas, but it was significant for only machine labour and capital expenses.

For pooled sample also the same trend was found, with the only exception of seed and fertilizer input exerted negative influence but none of them was significant at any acceptable level of significance. As we know, in the Cobb-Douglas production function, regression coefficients are equivalent to production elasticities, it could be noticed that the production elasticities of all the inputs were invariably less than unity, thus implying diminishing marginal productivity with respect to each of inputs.

Table 10.3 presents the geometric mean values of various inputs and output on per ha basis in both the areas. It could be seen that the average expenditure on inputs used on the farms at control areas were higher as compared to those at watershed areas, whereas, the gross value of output of watershed farms was much higher than that of the control area's farms.

Table 10.3: Geometric Mean Levels of Input and Output

(Rs./ha)

Sl.No.	*Input/Output*	*Watershed Area*	*Control Area*	*Pooled*
1.	Human labour (N)	3034.63	2925.85	2979.74
2.	Seed (S)	2207.75	2325.86	2266.04
3.	Fertilizer (F)	751.51	771.56	761.47
4.	Machine labour (M)	2671.16	2757.46	2713.97
4.	Capital expenses (K)	482.89	451.59	466.98
5.	Output (Y)	19442.82	14288.74	16667.74

The decomposition analysis was used to estimate the contribution of various sources to the productivity difference in terms of gross returns from crop enterprises between watershed beneficiaries and non-beneficiaries in the region and the results are presented in Table 10.4. The technological and input use differentials between the two areas together contributed to form the total differences of the order of 36 per cent, whereas, the technological component alone accounted for 31 per cent. This implies that with the present level of resource used by the farmers at control areas, returns could be increased by about 32 per cent if watershed programme implemented thereof. In other words, with no additional units of inputs, the existing level of production could be increased to a great extent by the implementation of watershed development programmes.

The contribution of the neutral technological component in the productivity difference was negative (-16.98 per cent) whereas the non-neutral technological component contributed positively (47.49 per cent) to the total difference in the gross value of output. The positive non-neutral technological component signifies that with the present level of input used in the control areas, the farmers could have increased the returns by 47 per cent on implementation of watershed development programmes provided that the efficiency levels of input use were held constant. The total contribution of differences in the levels of input use to gap in returns accounted for only 1 per cent, indicating that the returns from crops on the farms at control areas could be increased by 1 per cent, if the per hectare input use levels on these farms could be increased to the same level as on the watershed farms.

Table 10.4: Decomposition of differences in Returns from Crop Enterprises between Watershed and Control Areas

	Sources of Difference	*Per cent Contribution*	
		Sub-Total	*Total*
A.	Total observed difference		36.07
B.	Technical change		30.51
	I. Neutral technological difference	–16.98	
	II. Non-neutral technological difference	47.49	
C.	Due to difference in input use level		1.49
D.	Total estimated difference (due to all sources)		32.00

There was slight discrepancy between the observed (36.07 per cent) and the estimated (32.00 per cent) differences in the returns in watershed and control areas farms. This discrepancy was attributed to the random error term and exclusion of one of the important variables from the model, namely, the management input (Bisaliah, 1977; Umesh and Bisaliah, 1986; Lalwani, 1989 and Kiresur *et al.*, 1995). However, in the present case, since the discrepancy in question was of a very low order the results of this decomposition analysis could be considered to be satisfactory.

Conclusion

The analysis revealed that most of the inputs for crop cultivation were used at significantly higher rate in control villages, however, the gross output were significantly higher at watershed villages. The decomposition analysis carried out to dis-aggregate the effect of various factors which caused differences in output between watershed beneficiaries and non-beneficiaries revealed that watershed technology contributed mostly for the variation. The cumulative contribution of differences in the levels of input use to the productivity gap was negligible or even negative for the selected crops. This indicates that with no additional or even with lower inputs, the existing level of production could be increased to a great extent by the implementation of watershed development programmes. This shows the vital contribution of integrated watershed management interventions in mitigating the effects of drought-induced shocks on livelihoods. Therefore, the implementation of watershed development programme needs to be extended to all the un-treated villages for all-round development of people at marginalized areas.

Summary

Total change in production and productivity are the two important dimensions of benefits of watershed development programmes along with the conservation of land and water resources. To segregate the impact of various watershed based interventions on crop productivity a study was carried out in Bundelkhand region of Madhya Pradesh. Data were collected from 240 farmers' selected from eight watersheds and eight control villages in the region using a multi-stage random sampling technique. Analysis of data indicates that implementation of watershed development programmes led to significant differences in productivity of major crops in the sampled watersheds. Decomposition analysis of difference in returns from crop enterprises between watersheds and control villages indicates that the contribution of technological component was positive and higher than the contribution of input differentials which signifies that with the existing level of input use in the control areas, the farmers could have increased the returns from crops on implementation of watershed development programmes. This calls for a wider coverage of watershed development programmes in order to bring all the areas under land treatment activities for improving the productivity levels.

References

Bisaliah, S. (1977). Decomposition analysis of output change under new production technology in wheat farming: Some implications to returns on research investment. *Indian Journal of Agricultural Economics*, **32** (3): 193-201.

Inter-Ministerial Central Team (2008). *Report on Drought Mitigation Strategy for Bundelkhand Region of Uttar Pradesh and Madhya Pradesh*, Government of India.

Kerr, John, Pangare, G., Pangare, V.L. and George, P.J. (2000). *An Evaluation of Dryland Watershed Development Projects in India*. Environment and Production Technology Division (ETPD) Discussion Paper No.68, International Food Policy Research Institute (IFPRI), Washington. D.C.

Kiresur, V.R. and Ichangi, M. (2011). Socio-economic Impact of Bt Cotton – A Case Study of Karnataka. *Agricultural Economics Research Review*, **24**(1): 67-81.

Kiresur V., Pandey, R.K. and Mruthyunjaya (1995). Technological change in sorghum production: An economic study of Dharwad farms in Karnataka. *Indian Journal of Agricultural Economics*, **50** (2): 185-192.

Lalwani, M. (1989). Technological change in India's dairy farming sector: Distribution and decomposition of output gains. *Indian Journal of Agricultural Economics*, **44** (1): 55-66.

Reddy, V.R., Reddy, G., Galab, S., Soussan, J. and Baginski, S. (2004). Participatory Watershed Development in India: Can it Sustain Rural Livelihoods? *Development and Change*, **35**(2): 297-326.

Sharma, B.R. and Scott, C.A. (2005). Watershed Management Challenges: Introduction and Overview, In: *Watershed Management Challenges: Improving Productivity, Resources and Livelihoods*. Eds: B.R. Sharma, J.S. Samra, C.A. Scott and S.P. Wani, International Water Management Institute (IWMI) and International Crop Research Institute for Semi-arid Tropics (ICRISAT), Malhotra Publishing House, New Delhi. pp. 245-257.

Umesh, K.B. and Bisaliah, S.(1986). Productivity differential between small and large farms: An econometric study. *Asian Economic Review*, **28**(3): 17-33.

Chapter 11

Impact Assessment of the Farmers Trainings on Scaling-up of Water Productivity in Agriculture

Souvik Ghosh, Ashwani Kumar and Tanmaya Mohapatra

Introduction

Water scarcity has emerged as one of the challenging issues in recent years. So, it should be conserved, harvested and utilized in an efficient and profitable way in order to get more gain per drop of water. Present day agriculture and Indian farming community is facing multiple problems to maximize crop productivity. Inspite of successful research on new agricultural practices related to crop cultivation, the majority of farmers are not getting upper bound yield due to several reasons. One of the concerns that often sighted is scientific advice regarding crop cultivation does not reach to the farming community in a timely manner. Extension agency plays a major role in bridging this gap and communicate the latest technologies to the farmers (Kumar *and* Ratnakar, 2011). To increase the productivity of crop, the adoption of improved technology is a pre–requisite (Yadav *et al.*, 2011). Use of new or improved inputs and adoption of technology in agriculture are important in increasing farm productivity (Kher *et al.*, 2005). Adoption of technology depends on the required information received by the farmers from different sources as it is a critical input and as important as other key inputs such as credit, seeds, fertilizers and water. Different sources and channels of agriculture information can play important role to meet this requirement (Yadav *et al*, 2011). The adoption of an innovation depends upon its impact on the minds of potential adopters (Prasad and Katteppa, 2005). Farmers' attitudes determine adoption of improved technology (Chilonda and Van Huylenbroeck, 2011). Attitude is operationalised as the degree of positive or negative feeling of farmers towards the technologies communicated by extension services (Kumar *and* Ratnakar, 2011). Favourable attitude is essential for acceptance of any scientific innovation (Kanwat *et al.*, 2011).

Directorate of Water Management, Bhubaneswar conducted 69 farmers' trainings (each of seven days) in different districts of Odisha during the year 2007 to 2011. These trainings courses mainly focused in disseminating the scientific techniques to farmers those would enhance the water productivity in agriculture and improve livelihoods. Here water productivity is defined as the production obtained (Rs.) per unit volume of water (m^3). Therefore, enhanced water productivity includes application of all scientific techniques along with efficient and multiple uses of water in agriculture which as a whole increase the production. The present study was undertaken to assess the impact of these farmers trainings conducted during past years after a minimum gap of one year.

Methodology

The training effectiveness operationalised as the transfer effects of training: the extent to which farmers use their newly gained knowledge, skills and attitudes. Questionnaire surveys covering a sample of 25-30 per cent of the farmers trained during the farmers' trainings were carried out for impact assessment of the trainings after minimum time gap of one year. The data were collected from randomly selected 427 farmers undergone the trainings in sample of eight districts (Khurda, Puri, Cuttack, Nayagarh, Jajpur, Jagatsinghpur, Dhenkanal, Balasore) of Odisha through questionnaire survey for impact assessment.

Questionnaire was devised for collecting the information regarding the post training effects on the farmers. Social and personal information were collected and the feedbacks regarding awareness, attitude, skill and knowledge gain of the farmers were recorded on a three point continuum scale: low, medium, high with score as 1, 2 and 3, respectively. The data related to the adoption of the different scientific techniques with area of adoption were also collected.

The knowledge gain by the farmers was calculated by the following formula:

$$\text{Knowledge Gain (\%)} = \frac{\text{Mean Knowledge gain score of a techniques/practices/issues } (G_j)}{\text{Maximum Knowledge gain score for that techniques/practices/issues } (G_m)} \times 100$$

Where, $G_j = \Sigma G_{ij}/N$

Where

$i = 1,2, \ldots\ldots, N$

$j = 1,2, \ldots\ldots, K$

N = total number of respondents=427

K = total number of technologies/practices/issues = 28

G_m = Maximum Knowledge gain score = 3

The adoption of the techniques by the farmers was calculated by

$$\text{Extent of adoption } (\%) = \frac{\text{Number of farmers adopted a technique/practice} (A_t)}{\text{Total numbers of farmers surveyed}} \times 100$$

Where t = 1, 2,....., m

m=Total number of techniques/practices=25

The responses of the sampled farmer-trainees were analyzed using appropriate statistical techniques.

Result

Profile of the Farmers Trainees

The profiles of the farmers are given in Table 11.1. Most of the respondents (54.70 per cent) were under middle age (40-60 yrs) group. Relatively less number of young farmers calls for the need of encouraging farm youths to be actively involved in agriculture and to adopt scientific technologies being more educated and positive mindset than old farmers (CIMMYT, 1993). Around 48 per cent of farmers were below poverty lines. Majority of the farmers (94.50 per cent) were found to be undergraduate and around 47 per cent farmers had no experience of participation in trainings during past years. So, it reiterates the need of the training to these farmers to make them aware about the modern scientific technologies for improved agriculture with more productivity and farm income. The education through training is expected to enhance the decision making and the adoption of agricultural technologies (Cavane, 2011). Adoption of technology is influenced by physical, socio-economic and psychological factors like agro-ecological conditions, age, family size, education, how-to-knowledge, source of information and farmer's attitude towards the technology (Neupane *et al.*, 2002 and Rogers, 2003).

Impact of Training

Impact of training was realized through analyses of improvement in awareness, attitude and skill of farmers undergone the training. From Figure 11.1, it is revealed that overall improvement in awareness, attitude and skill of the farmers are found to be medium and in the range of 65 per cent to 70 per cent which showed that the trainings had positive effects on the farmers about the modern scientific techniques for scaling-up of water productivity in agriculture. These results are in conformity with findings of Das and Sharma (1998) who also found that training programmes contributed significantly in improvement of respondent's knowledge about scientific practice.

Table 11.1: Profile of the Farmers Undergone Trainings

N=427

Sl.No.	Variables	Category	Frequency	Percentage
1.	Gender	Male	358	83.84
		Female	69	16.16
2.	Age	Young (0-40 years)	134	31.49
		Middle (40-60 years)	234	54.70
		Old (60-90 years)	59	13.81
3.	Economic Status	BPL	205	47.91
		APL	222	52.09
4.	Education	Illiterate	3	0.74
		Functional literate	44	10.32
		Primary	113	26.54
		Secondary	167	39.07
		Higher Secondary	77	17.94
		Diploma/Degree	23	5.41
5.	Training attended during past years	Nil	199	46.56
		1-2 trainings	183	42.98
		3-4 trainings	34	7.99
		4-5 trainings	4	0.83
		More than 5 trainings	7	1.65

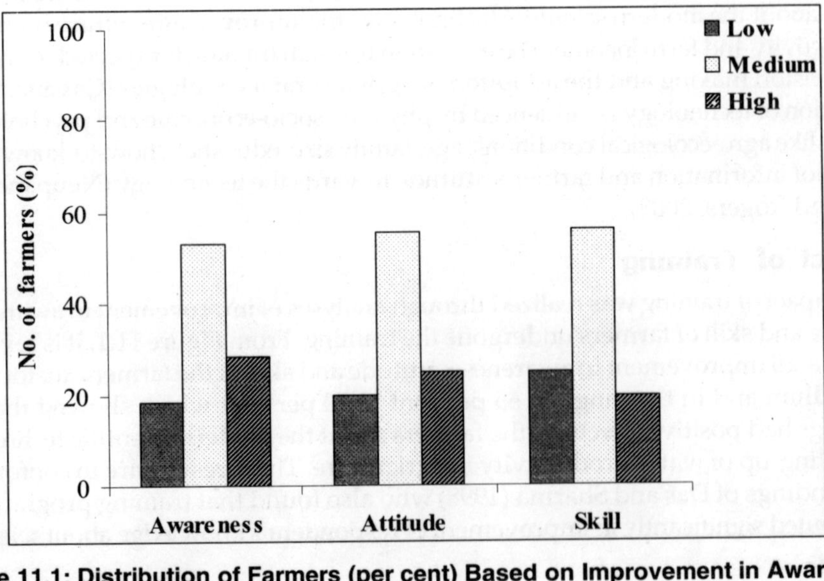

Figure 11.1: Distribution of Farmers (per cent) Based on Improvement in Awareness, Attitude and Skill Towards Scaling-Up of Water Productivity in Agriculture.

Table 11.2 shows the mean value of improvement in awareness, attitude and skill of the farmers of different districts. It was found that awareness of the farmers of Balasore district was higher in comparison to other districts. The attitude change was found more for the farmers of Jagatsinghpur district while the farmers of Nayagarh showed maximum skill development. The variation (standard deviation) in improvement of all three criteria is found to be almost same; however, the improvement in awareness was relatively more as compared to attitude and skill.

Table 11.2: Improvement in Awareness, Attitude and Skill of Farmers of different Districts in Odisha Undergone Training

Sl.No.	*Name of the District*	*Awareness*		*Attitude*		*Skill*	
		Mean	*SD*	*Mean*	*SD*	*Mean*	*SD*
1.	Khurda (118)	2.01	0.59	2.07	0.56	2.13	0.60
2.	Puri (54)	1.76	0.43	1.33	0.51	1.35	0.65
3.	Cuttack (45)	2.31	0.67	2.20	0.63	2.16	0.64
4.	Nayagarh (15)	2.33	0.49	2.07	0.73	2.20	0.56
5.	Jajpur (20)	2.05	0.51	1.80	0.52	1.85	0.59
6.	Jagatsinghpur (65)	1.98	0.85	2.52	0.64	1.92	0.74
7.	Dhenkanal (80)	2.18	0.73	2.04	0.64	1.90	0.57
8.	Balasore (30)	2.73	0.45	2.17	0.38	2.17	0.38
	Overall (427)	2.10 (70.10%)	0.68	2.05 (68.27%)	0.67	1.95 (64.91%)	0.66

Note: Figure in the parenthesis indicate number of farmers-respondents.

Data presented in Table 11.3 revealed that the overall knowledge gain on different scientific techniques was 62 per cent which showed that the trainings had helped farmers in increasing their level of knowledge regarding new scientific technologies for improved agriculture which in turn helped them towards adoption of these technologies gaining more production and farm income. Farmers acquired maximum knowledge on different techniques/issues *viz.* growing pulse crops through residual soil moisture utilization during *rabi* season (71 per cent), horticultural crops cultivation: vegetables, fruit crops (71 per cent), farm implements and their uses/ farm mechanization (70 per cent), growing more rice with less water through SRI method (70 per cent), rainwater conservation/harvesting techniques (69 per cent), mushroom cultivation (68 per cent), use of organic fertilizers (bio fertilizers, vermi-compost, etc.) (68 per cent), judicious use/optimum dose of fertilizers (67 per cent), livestock rearing (cattle/goat/sheep) (66 per cent), multiple uses of conserved rainwater (65 per cent), groundwater utilization through dugwell, borewell, tubewell, etc (65 per cent); crop diversification (65 per cent), availing benefits as beneficiary farmer under different schemes (65 per cent), plant protection measures: insect and disease management in major crops (64 per cent), water-saving irrigation methods instead of flooding method of irrigation (63 per cent), economic use of waterlogged areas through cultivation of aquatic crops, fish farming (62 per cent) and conjunctive use of groundwater and surface water (61 per cent).

Table 11.3: Knowledge Gain and Adoption of the Techniques/Practices by the Farmers Trained through One-Week Farmers Training Programmes

Sl.No.	*Issues/Techniques/Practices Discussed during the Training Programmes*	*N = 427*		
		No. of Farmers Adopted (Per cent)	*Adopted Area in ha.*	*Knowledge Gain (per cent)*
1.	Rainwater conservation/harvesting techniques	64	117	69
2.	Multiple uses of conserved rainwater	53	94	65
3.	Ground water utilization through dugwell, borewell, tubewell, etc	51	74	65
4.	Micro level water resources development like tank, sub-surface water harvesting, well, farm pond, etc	29	30	59
5.	Water conservation measures for rainfed areas/watersheds including rubber dam as check dam	9	4	53
6.	Water-saving irrigation methods instead of flooding method of irrigation	34	57	63
7.	Conjunctive use of ground water and surface water	39	68	61
8.	Pressurized irrigation systems (drip, sprinkler)	13	9	55
9.	Farm implements and their uses/farm mechanization	79	154	70
10.	Growing more rice with less water through SRI method	54	91	70
11.	Growing pulse crops through residual soil moisture utilization during *rabi* season	76	124	71
12.	Crop diversification	49	66	65
13.	Horticultural crops cultivation: vegetables, fruit crops	65	60	71
14.	Cultivation of high value crops like medicinal plants, cash crops	27	26	53
15.	Economic use of waterlogged areas through cultivation of aquatic crops, fish farming	35	33	62
16.	Rice-fish integration	16	7	52
17.	Mushroom cultivation	22	7	68
18.	Integrated farming system approach	31	30	58

Contd...

Table 11.3–*Contd...*

Sl.No.	Issues/Techniques/Practices Discussed during the Training Programmes	N = 427		
		No. of Farmers Adopted (Per cent)	Adopted Area in ha.	Knowledge Gain (per cent)
19.	Soil testing measures and important parameters	37	63	55
20.	Soil nutrient management (nutrient deficiency symptoms and corrective measures)	37	62	52
21.	Judicious use/optimum dose of fertilizers	73	117	67
22.	Use cf organic fertilizers (bio fertilizers, vermi-compost, etc.)	72	102	68
23.	Plant protection measures: insect and disease management in major crops	73	93	64
24.	Poultry farming	22	4	58
25.	Livestock rearing (cattle/goat/sheep)	62	7	66
26.	Group approach/participatory approach for self empowerment			59
27.	Financial issues: agricultural loans/crop loans/crop insurance schemes			59
28.	Availing benefits as beneficiary farmer under different schemes			65
	Overall	45	1500	62

It is evident that the respondents-farmers have put the scientific techniques (25 techniques) into practices in a total area of 1500 ha. Figure 2.2 depicts adoption of different technologies. Maximum adoption by the farmers was observed for the techniques/practices like farm implements and their uses/farm mechanization (79 per cent) in an area of 154 ha., growing pulse crops through residual soil moisture utilization during *rabi* season (76 per cent) in an area of 124 ha., judicious use/optimum dose of fertilizers (73 per cent) in an area of 117 ha., plant protection measures: insect and disease management in major crops (73 per cent) in an area of 93 ha., use of organic fertilizers (bio-fertilizers, vermi-compost, etc.) (72 per cent) in an area of 102 ha., horticultural crops cultivation: vegetables, fruit crops (65 per cent) in an area of 60 ha., rainwater conservation/harvesting techniques (64 per cent) in area of 117 ha., growing more rice with less water through SRI method (54 per cent) in area of 91 ha., multiple uses of conserved rainwater (53 per cent) in area of 94 ha. and groundwater utilization through dugwell, borewell, tubewell, etc (51 per cent) in an area of 74 ha. From the results, it can be concluded that the trainings had motivated the farmers for adoption of scientific techniques in their field for gaining more production and income. These results are in conformity with findings of Kaliba *et al.*, 2000; Abebaw and Belay, 2001; Neupane *et al.*, 2002; Roger, 2003; Hintz *et al*, 2003 and Ahire *et al.*, 2007 in which they reported that knowledge, source of information, mass media exposure, role of extension in dissemination of improved technologies influenced the adoption of improved technologies.

Conclusion

The farmers undergone trainings hold positive approach towards the adoption of modern scientific techniques in agriculture for gaining more production and income. The trainings were conducted in location specific manner addressing location specific water management problems in agriculture. Thus the farmers gained not only the knowledge about the techniques for enhancing water productivity in agriculture but also developed their skill with changed positive attitude which are reflected in adoption of the technologies by substantial number of farmers. On–farm trainings are found to have more immediate impact; therefore, the Krishi Vigyan Kendras and State line departments need to emphasize more on location specific on-farm trainings of the farmers rather than only on-station routine farmers trainings.

Summary

The present study was attempted to know the impact of the farmers training programmes conducted in eight districts of Odisha during the year 2007-11.These training programmes each of seven days duration were organized by Directorate of Water Management, Bhubaneswar with major theme of the trainings "scaling-up of water productivity in agriculture for livelihoods". To assess the impact of trainings, questionnaire surveys covering 25-30 per cent of farmers trained during the training programmes were carried out after a gap of minimum one year period. The findings implied that farmers showed significant improvement in their overall awareness (70.10 per cent), attitude (68.27 per cent) and skill (64.91 per cent). The overall knowledge gain of the farmers on different scientific techniques for enhancing water productivity in agriculture was found to be significant and the adoption of these

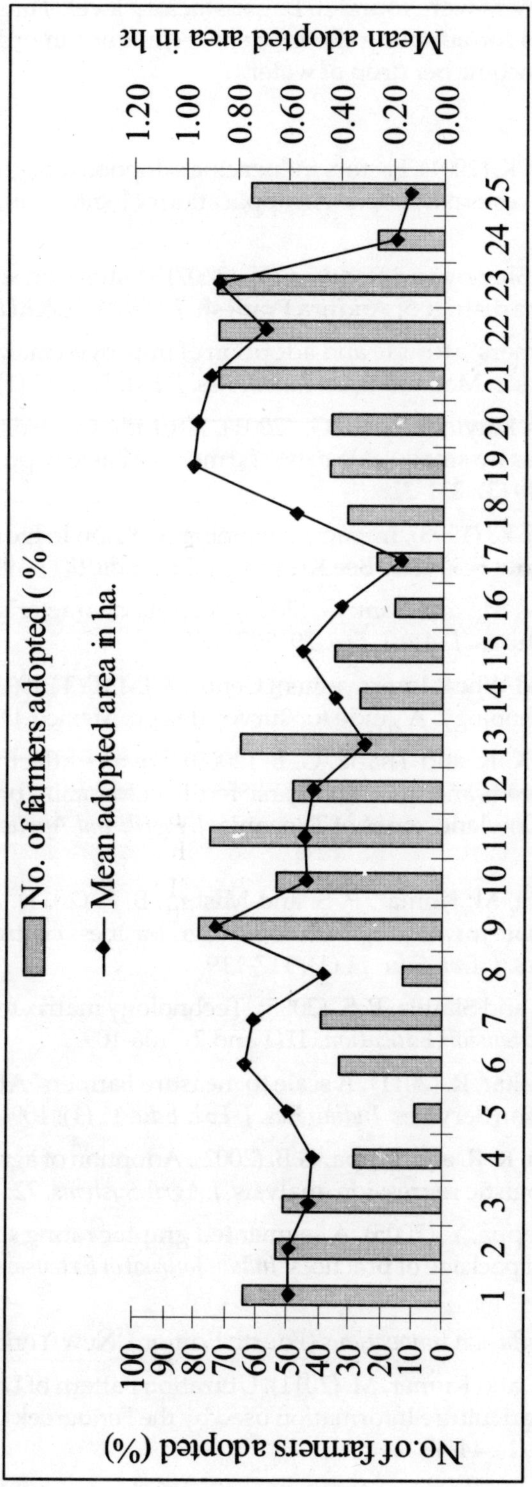

Figure 11.2: Techniques Adopted by the Farmers (per cent) with Mean Area of Adoption.

techniques by the farmers were found to be satisfactory level. Thus, trainings had motivated the farmers for adopting the modern techniques in order to get more production and farm income per drop of water.

References

Abebaw, D., and Belay, K. (2001). Factors influencing adoption of high yielding maize varieties in Southwestern Ethiopia: An application of logit. *Quarterly J. Int. Agril., 40*(2): 149-167.

Ahire, I. M.; Sandhya, Shenoy and Reddy, C.V. (2007). A study on adoption level of growers of Krishna district of Andhra Pradesh. *J. Res. ANGRAU*, 35(1): 73-76.

Cavane, E. (2011). Farmers' attitude and adoption of improved maize varieties and chemical fertilizers in Mozambique. *Indian Res. J. Ext. Edu.* 11 (1): 1-6

Chilonda, P., and Van Huylenbroeck, G. (2001). Attitude towards and uptake of veterinary services by small-scale cattle farmers in Eastern province Zambia. *Outlook on Agri.*, 30 (3): 231-218.

Das, P.K. and Sharma J. K. (1998), Impact of Training on Knowledge and Perception of Rural Youth about Scientific Bee Keeping, *J.Extn.Edu.* 9(1): 1957-1962.

Hintze, L. H., Renkow, M., and Sain, G. (2003). Variety characteristics and maize adoption in Honduras. *J. Agril. Eco.* 29: 307-317.

International Maize and Wheat Improvement Centre (CIMMYT).1993. The adoption of agricultural technology. A guide for survey design. Mexico, D. F.: CIMMYT.

Kaliba, R.P., Sharma, K. R. and Thapa, G. B. (2000). Factors affecting adoption of improved maize seeds and use of inorganic fertilizer for maize production in the intermediate and lowland zones of Tanzania. *J. Agril. and Applied Eco.* 32(1): 35-47.

Kanwat, M.; Chargotra, M; Kumar, P. S. and Mishra, B. P. (2011). Attitude of the agricultural graduate towards agri-clinic and agri business centres in Arunachal Pradesh. *Indian Res. J. Ext. Edu.* 11 (1): 117-119

Kher, S. K.; Patel, R. B. and Slathia, P. S. (2005). Technology matrix for rice growers. *Indian Journal of Extension Education.* 41(1 and 2): 106-109.

Kumar, P. G. and Ratnakar, R. (2011). A scale to measure Farmers' Attitude towards ICT-based Extension Services. *Indian Res. J. Ext. Edu.* 11 (1): 109-112

Neupane, R.P., Sharma, K. R. and Thapa, G.B. (2002). Adoption of agroforestry in the hills of Nepal: a logistic regression analysis. *J.Agril.Systems.* 72: 177-196.

Prasad, V. L. and Katteppa, Y. (2005). A segmented graphic rating scale to measure appropriateness of package of practices. *Indian Journal of Extension Education.*41(1 and 2): 91-94.

Rogers, E.M. (2003). *Diffusion Innovations* (Fourth Edition). New York: Free Press.

Yadav, B.S.; Khan, I. M. and Kumar, M. (2011). Utlization Pattern of Different sources and channels of Agriculture Information used by the Fenugreek Growers. *Indian Res. J. Ext. Edu.* 11 (1): 44-49.

Chapter 12

Integrated Crop-Livestock Farming Systems: A Strategy for Resource Conservation and Environmental Sustainability

Vinod Gupta, Pradeep Kumar Rai and K.S. Risam

Introduction

In recent years, food security, livelihood security, water security as well as natural resources conservation and environment protection have emerged as major issues worldwide. Developing countries struggling to deal with these issues and also have to contend with the dual burden of climate change and globalization. It has been accepted by everyone across the globe that sustainable development is the only way to promote rational utilization of resources and environmental protection without hampering economic growth. Developing countries around the world are promoting sustainable development through sustainable agricultural practices which will help them in addressing socio-economic as well as environmental issues simultaneously. Within the broad concept of sustainable agriculture "Integrated Farming Systems" hold special position as in this system nothing is wasted, the by-product of one system becomes the input for other. Integrated farming is an integrated approach to farming as compared to existing monoculture approaches. It refers to agricultural systems that integrate livestock and crop production. Moreover, the system help poor small farmers, who have very small land holding for crop production and a few heads of livestock to diversify farm production, increase cash income, improve quality and quantity of food produced and exploitation of unutilized resources. Population growth, urbanization and income growth are fuelling a substantial increase in the demand for food of animal origin, while also aggravating the competition between crops and livestock (increasing cropping areas and reducing rangelands). The livestock revolution is stretching the capacity of existing production, but it is also exacerbating environmental problems. Therefore, while it is necessary to satisfy consumer demand, improve nutrition and direct income growth opportunities to those who need those most, it is also necessary to alleviate environmental stress (Delgado *et al.*, 1999).

An integrated farming system consists of a range of resource-saving practices that aim to achieve acceptable profits and high and sustained production levels, while minimizing the negative effects of intensive farming and preserving the environment. Based on the principle of enhancing natural biological processes above and below the ground, the integrated system represents a winning combination that

1. Reduces erosion
2. Increases crop yields, soil biological activity and nutrient recycling
3. Intensifies land use, improving profits
4. Can therefore help reduce poverty and malnutrition and strengthen environmental sustainability.

Livestock and crop production systems are an integral part of one another (Kallah and Adamu 1988). Crop residues provide fodder for livestock (Van Raay and de Leeuw 1971; Al hassan *et al.*, 1983) while, occasionally, grain provides supplementary feed for productive animals. Animals improve soil fertility through manure and urine deposition and animal power for farm operations and transport. Sale of animals sometimes provides cash for farm labor and agricultural inputs. There are several examples of completely integrated crop–livestock production systems where sustainable increases in both crop and livestock production have been achieved after considerable periods (30–40 years) of continuous cropping without resulting in land degradation. Some of them are the close settled zone (CSZ) of Kano in northern Nigeria (Harris 1995), Banamba in Central Mali (Abou Berthe, personal communication), and Batalay in southern Chad (K.N. Ngwara, personal communication). The key success to these farming systems is effective crop–livestock integration involving the recycling of nutrients within the system. A particular challenge facing farmers is to minimize nutrient losses through good management (Powell and Valentin 1997); improved feed production, quality, availability, and more efficient feeding systems; new ways to capture and conserve nutrients excreted by livestock; improved manure spreading techniques; and cropping systems that reduce nutrient losses and can improve livestock impacts on the soil environment.

The farming system is essentially cyclic (organic resources – livestock – land – crops). Therefore, management decisions related to one component may affect the others.Using crop residues more rationally is an important route out of poverty. For resource-poor farmers, the correct management of crop residues, together with an optimal allocation of scarce resources, leads to sustainable production.Combining ecological sustainability and economic viability, the integrated livestock-farming system maintains and improves agricultural productivity while also reducing negative environmental impacts. Some lessons learned and recommendations

- ☆ The maintenance of an integrated crop livestock system is dependent on the availability of adequate nutrients to sustain animals and plants and to maintain soil fertility. Animal manure alone cannot meet crop requirements, even if it does contain the kind of nutrients needed. This is because of its relatively low nutrient density and the limited quantity available to small-scale farmers. Alternative sources for the nutrients need to be found.
- ☆ Growing fodder legumes and using them as a supplement to crop residue is the most practical and cost-effective method for improving the nutritional value of crop residues. This combination is also effective in reducing weight loss in animals, particularly during dry periods;
- ☆ Given their traditional knowledge and experience, local farmers are perfectly able to apply an integrated system. In practice, however, relatively few adopt this system, mainly because they have limited access to credit, technology and knowledge. The crop-pasture rotation system is complex and requires a substantial capital outlay for machinery and implements. Associations of grain and livestock producers are useful for filling these gaps and can promote the adoption of a crop livestock system;
- ☆ Better livestock management is needed to safeguard water. Livestock water demand includes water for drinking and for feed production and processing. Livestock also have an impact on water, contaminating it with manure and urine. All of these aspects need to be given due consideration.
- ☆ Intensification of agriculture through appropriate incorporation of small livestock has the potential to decrease the land needed for agricultural production and relieve the pressure on forests.

Advantages

Economic analysis of different farming systems (one hectare of irrigated land or 1.5 ha of un-irrigated land) indicated that under irrigated conditions, mixed farming with crossbred cows yielded the highest net profit, followed by mixed farming with buffalo, and arable farming. Mixed farming with Hariana cows made a loss (Singh *et al.*, 1993). Comparative productivity and economies of dairy enterprises (mixed farming with three crossbred cows on one hectare of canal irrigated land versus mixed farming with three *Murrah* buffalo) indicated that mixed farming with crossbred cows under canal-irrigated conditions was more efficient for the utilisation of land, capital, inputs and the labour resources of the farmer (Kumar *et al.*, 1994). Baseline surveys in Gujarat India, indicated that around 75 per cent of rural households kept cattle in the face of under-employment. More particularly, the farm surveys showed that cattle kept mainly for milk, contributed 32 per cent and 20 per cent for tribal and non-tribal ethnic groups respectively (Patil and Udo 1997). By comparison to cows and buffaloes, lactating goats contributed between 54-68.9 per cent to total farm income through the sale of milk (Deoghare and Bhattacharyya 1993; 1994; Deoghare and Sood 1994). The significance of milk production from goats and the links to food security and

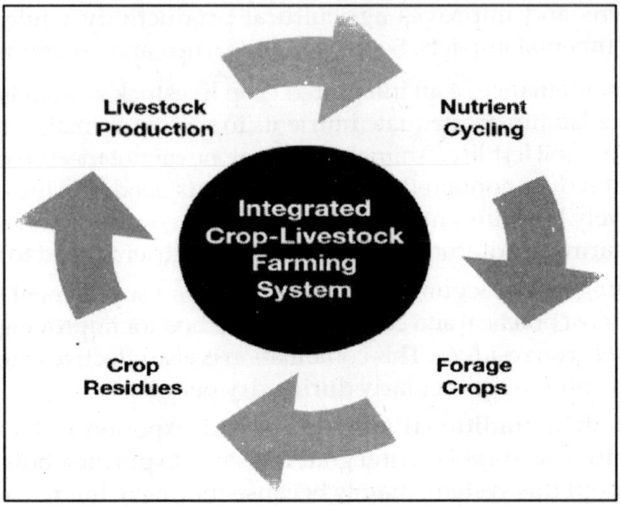

Figure 12.1: Integrated Crop-Livestock Farming System: Key Aspects.

livelihoods of the poor has recently been reviewed (Devendra 1996). In an integrated system, livestock and crops are produced within a coordinated framework. (Van Keulen and Schiere, 2004). The waste products of one component serve as a resource for the other. For example, manure is used to enhance crop production; crop residues and by-products feed the animals, supplementing often inadequate feed supplies, thus contributing to improved animal nutrition and productivity.

The result of this cyclical combination is the mixed farming system, which exists in many forms and represents the largest category of livestock systems in the world in terms of animal numbers, productivity and the number of people it services. (Van Keulen and Schiere, 2004).Animals play key and multiple roles in the functioning of the farm, and not only because they provide livestock products (meat, milk, eggs, wool, and hides) or can be converted into prompt cash in times of need. Animals transform plant energy into useful work: animal power is used for ploughing, transport and in activities such as milling, logging, road construction, marketing, and water lifting for irrigation. Animals also provide manure and other types of animal waste. Excreta have two crucial roles in the overall sustainability of the system:

(a) Improving Nutrient Cycling

Excreta contain several nutrients (including nitrogen, phosphorus and potassium) and organic matter, which are important for maintaining soil structure and fertility. Through its use, production is increased while the risk of soil degradation is reduced.

(b) Providing Energy

Excreta is the basis for the production of biogas and energy for household use (*e.g.* cooking, lighting) or for rural industries (*e.g.* powering mills and water pumps). Fuel in the form of biogas or dung cakes can replace charcoal and wood.Crop residues represent the other pillar on which the equilibrium of this system rests. They are fibrous by-products that result from the cultivation of cereals, pulses, oil plants, roots and tubers. They are a valuable, low-cost feed resource for animal production, and are consequently the major source of nutrients for livestock in developing countries. The overall benefits of crop-livestock integration can be summarized as follows:

- Agronomic, through the retrieval and maintenance of the soil productive capacity;
- Economic, through product diversification and higher yields and quality at less cost;
- Ecological, through the reduction of crop pests (less pesticide use and better soil erosion control); and
- Social, through the reduction of rural urban migration and the creation of new job opportunities in rural areas.

This system has other specific advantages

- It helps improve and conserve the productive capacities of soils, with physical, chemical and biological soil recuperation. Animals play an important role in harvesting and relocating nutrients, significantly improving soil fertility and crop yields.
- It is quick, efficient and economically viable because grain crops can be produced in four to six months, and pasture formation after cropping is rapid and inexpensive.
- It helps increase profits by reducing production costs. Poor farmers can use fertilizer from livestock operations, especially when rising petroleum prices make chemical fertilizers unaffordable.
- It results in greater soil water storage capacity, mainly because of biological aeration and the increase in the level of organic matter.
- It provides diversified income sources, guaranteeing a buffer against trade, price and climate fluctuations.

Overall concept of partnerships and interactions for enhancing crop-livestock interactions for sustainable development production intensification.

The key advantage of crop-livestock production systems is that livestock can be fed on crop residues and other products that would otherwise pose a major waste disposal problem. For example, livestock can be fed on straw, damaged fruits, grains and household wastes (Fakoya, 2002). Integration of livestock and crop allow nutrients to be recycled more effectively on the farm. Manure itself is a valuable fertilizer containing 8 kg of nitrogen, 4kg of phosphorus and 16 kg of potassium to the tone (FAO, 1999). Adding manure to the soil not only fertilizes it but also improved its structures and water retention capacity (ILCA, 1988). FAO (1996) opined that where livestock are used to graze, the vegetation under plantations of coconut, oil palm and rubber, as in Malaysia, the cost of weed control can be dramatically reduced, sometimes by as much as 40 per cent. In Colombia sheep are sometimes used to control weeds in sugarcane. Draught animal power is widely used for cultivation,

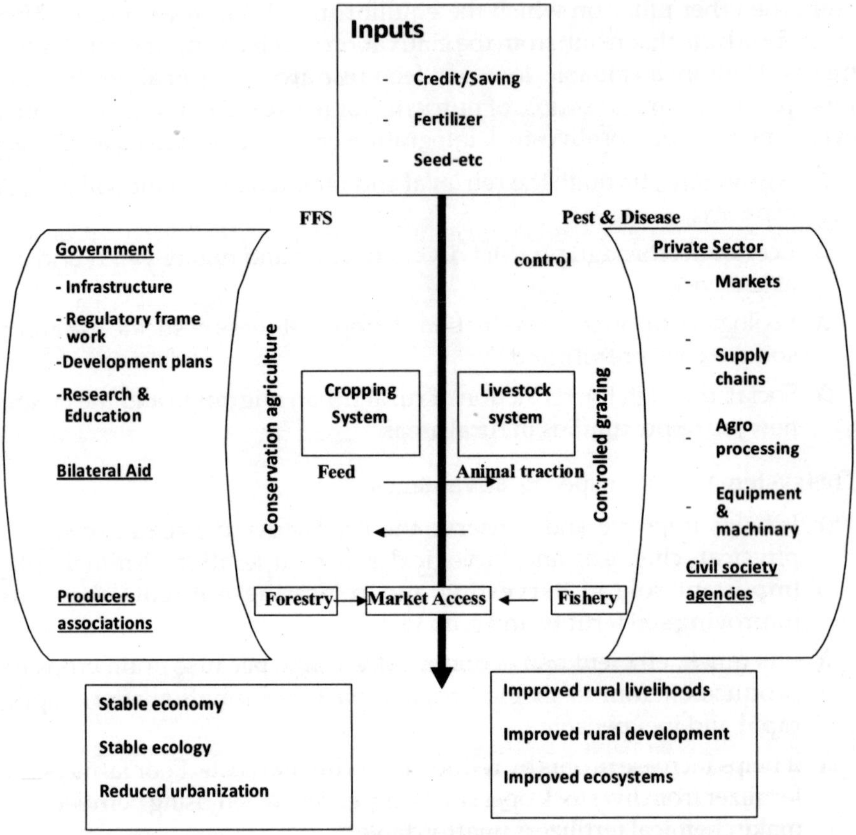

Figure 12.2

transportation, water lifting and powering food processing equipment. Using draught animal reduces the need for foreign exchange to buy expensive tractors and fuel (Jahnke, 1992). According to International food security treat campaign (1984) it was estimated that 52 per cent of the cultivated area in developing countries excluding China is farmed exclusively with draught animal, animal traction, bringing heavy but potentially very productive soil into production. According to FAO (1997) cow dung is highly valued for used for cooking and heating in many countries. Alternatively, 25kg of fresh cow dung makes on cubic metre of biogas, which can be used to provide energy for light, heat or motive power.

Constraints

- ✩ Nutritional values of crop residues are generally low in digestibility and protein content. Improving intake and digestibility of crop residues by physical and chemical treatments is technically possible but not feasible for poor small farmers because they require machinery and chemicals that are expensive or not readily available (Keftasa and International Livestock Centre for Africa, 1988).

- ☆ Crop residues are primarily soil regenerators, but too often they are either disregarded or misapplied.
- ☆ Intensive recycling can cause nutrient losses.
- ☆ If manure nutrient use efficiencies are not improved or properly applied, the import of production and transportation, and the surpluses lost in the environment.
- ☆ Farmers prefer to use chemical fertilizer instead of manure because it acts faster and is easier to use.
- ☆ Resource investments are required to improve intake and digestibility of crop residues.

Mixed farms are prone to using more manure than crop farms do. Manure transportation is an important factor affecting manure use.

Challenge

Develop strategies and promote crop livestock synergies and interactions that aim to

(a) Integrate crops and livestock effectively with careful land use;

(b) Raise the productivity of specific mixed crop-livestock systems;

(c) Facilitate expansion of food production; and

(d) Simultaneously safeguard the environment with prudent and efficient use of natural resources.

Devise measures (for instance, facilitating large-scale dissemination of bio-digesters) to implement a more efficient use of biomass, reducing pressures on natural resources; and develop a sustainable livestock manure management system to control environmental losses and contaminant spreading.

Opportunities

- ☆ Intensification of agriculture which is currently occurring in most farming systems favours crop–livestock integration.
- ☆ Poor soil fertility, unavailability or increases in prices of fertilizers, and labour shortages, have forced farmers to rely on alternatives such as manure and traction.
- ☆ Farmers can grow crop in the wet season and engage in livestock enterprises in the dry season.
- ☆ Livestock enterprises are more lucrative than crop farming so it is advantageous to integrate livestock into farm activities.
- ☆ Many indigenous, emerging, and developed technologies are available to support sustainable crop–livestock integration. These include improved cereal and grain legume varieties, cropping systems, weed and nutrient management strategies, the eradication of most livestock diseases, and the development of modeling and all-year-round feed packages for animals.

Systems modeling and GIS applications, combined with socioeconomic information, such as identification of areas where market driven intensification is ongoing and impact is likely to be greater (Ndubuisi *et al.*, 1998), should have a role in optimizing the targeting of new technologies. For instance, an approach needs to be included encompassing the "whole system" while aspects including the determinants of farmer decision-making and opportunities for using technologies (such as improved varieties, cropping systems, better ruminant nutrition, weed and nutrient management) to enable existing sustainable systems to address the challenges of increased intensification need to be included.Establishing effective input (*e.g.*, fertilizer) and support services (*e.g.*, veterinary delivery systems) and establishing infrastructure (*e.g.*, roads, processing, and marketing facilities); are important. An appropriate strategy would be to select technologies with the highest potential impact from the above areas, in order to form a holistic package for testing and dissemination with a view to maximizing total productivity at the farm level. This holistic approach should be supported by socioeconomic information such as determinants of farmers' decision making for certain technologies as well as the development of tools to assess whole farm impacts of new interventions.

Conclusion

Sustainable development is the only way to promote rational utilization of resources and environmental protection without hampering economic growth and integrated Farming Systems hold special position as in this system nothing is wasted, the by-product of one system becomes the input for other. India has a considerable livestock, poultry population and crop wastes. All efforts have to be mobilised to reclaim the resources and to put them to use effectively. Suitable technology has to be developed for the treatment of wastes and their all round effective utilisation, so that, it can help in to reducing the poverty and malnutrition and strengthen environmental sustainability.The increase in demand for livestock products presents opportunities for small farmers who can increase livestock production and benefit from related income. (Delgado *et al.*,1999). However, in terms of environmental impact, the growing number of livestock and the increase in livestock processing can have a negative impact on natural resources unless actions are taken to identify farming practices that are economically and ecologically sustainable. The highly improved integrated crop-livestock system can guarantee more sustainable production and therefore constitutes a valid new approach. Experience in the use of this system has shown that adopting sustainable management practices can improve production while preserving the environment.Residues, wastes and by-products of each component serve as resources for the others. Poor farmers have the traditional knowledge needed to integrate livestock and crop production, but because of their limited access to knowledge, assets and inputs, relatively few adopt an integrated system.

The challenge for development practitioners is to ensure that poor small farmers can increase the productivity of traditional farming systems, adopting an effective integrated system that produces usable biomass while conserving natural resources, and can therefore be sustainable in the long term. Within this framework some key questions discussions are:

- ☆ How can livestock production increase to meet the growing demand for livestock products, using methods that the resource base can sustain?
- ☆ Do the strategies devised for raising productivity in integrated crop-livestock systems take into account the stage of development of the target population with respect to the nature of crop-livestock interactions?
- ☆ Do the farmers concerned have the right skills, knowledge, capital and technology to set up this system?
- ☆ Are the roles and responsibilities of men and women given sufficient consideration?
- ☆ How can additional needed nutrients be obtained? And, can the productivity of the system be increased without stressing the environment?
- ☆ Is enough good-quality feed available to sustain animals, especially during the dry season?
- ☆ Are nutrients that are relocated from grazing areas to croplands efficiently recovered?
- ☆ Are the different components of the farming system (crop, livestock and, eventually, fish) efficiently integrated?

Therefore, a lot of extension attention is however required to educate farmers on utilization of crop-livestock production systems.

Summary

Conventional agriculture is known to cause soil and pasture degradation because it involves intensive tillage, in particular if practised in areas of marginal productivity. An integrated crop-livestock farming system represents a key solution for enhancing livestock production and safeguarding the environment through prudent and efficient resource use. The increasing pressure on land and the growing demand for livestock products makes it more and more important to ensure the effective use of feed resources, including crop residues. An integrated farming system consists of a range of resource-saving practices that aim to achieve acceptable profits and high and sustained production levels, while minimizing the negative effects of intensive farming and preserving the environment. Based on the principle of enhancing natural biological processes above and below the ground, the integrated system is the combination that (a) reduces erosion; (b) increases crop yields, soil biological activity and nutrient recycling; (c) intensifies land use, improving profits; and (d) can therefore help reduce poverty and malnutrition and strengthen environmental sustainability. The waste products of one component serve as a resource for the other. For example, manure is used to enhance crop production; crop residues and by-products feed the animals, supplementing often inadequate feed supplies, thus contributing to improved animal nutrition and productivity. Integrating crops and livestock serves primarily to minimize risk and not to recycle resources. In an integrated system, crops and livestock interact to create a synergy, with recycling allowing the maximum use of available resources. Crop residues can be used for animal feed, while livestock and livestock by-product production and processing can enhance agricultural productivity by intensifying nutrients that improve soil fertility, reducing the use of chemical fertilizers. A high integration of crops and livestock is often considered as a step forward, but small farmers need to have sufficient access to knowledge, assets and inputs to manage this system in a way that is economically and environmentally sustainable over the long term.

References

Alhassan, W.S., O.W. Ehoche, I.F. Adu, and A.T. Obilana. 1983. Crop residue potential of agricultural development projects. (a) Chemical composition of crop residues. NAPRI Annual Report, Shika, Zaria, Nigeria.

Deoghare P.K. and Bhattacharyya N.K. 1993. Economic analysis of goat rearing in Mathura district of Uttar Pradesh. *Indian Journal of Animal Sciences* 63:439-444.

Deoghare P.K. and Bhattacharyya N.K. 1994. Economics of Jamunapari goat rearing in Etawah district of Uttar Pradesh. *Indian Journal of Animal Sciences* 64:1390-1393.

Deoghare P.K. and Sood S.B. 1994. Income and employment potential of goat rearing on farms in the rural households of Mathura district of Uttar Pradesh. *Indian Journal of Animal Sciences* 64:295-300.

Devendra C. 1996. Opportunities for increasing the economic contribution of small ruminants in Asia. In: Jambre L.F. and Knox M.R. (eds), *Sustainable Parasite Control in Small Ruminants.* ACIAR Proceedings 74:27-32. ACIAR (Australian Centre for International Agricultural Research), Canberra, Australia.

Delgado, C., M. Rosegrant, H. Steinfeld, S. Ehui and C. Courbois. 1999. Livestock to 2020: The Next Food Revolution. Food, Agriculture and the Environment Discussion Paper 28, Washington, D.C.International Food Policy Research Institute.

Fakoya E.O. 2002. "Assessment of livestock production systems based on crop residues and legumes in humid zones of Nigeria." *Proceeding of the Nigeria society for Animal Production 27th Annual NASP Conference* Akure, Pp. 374 – 376.

FAO. 1996. Food and Agricultural Organization 1996. World Development Report Paper No. II Rome Italy. Pp. 51-54.

FAO. 1997 "Food and Agricultural Organization integrating crops and livestock in West Africa." *Animal Production Health Paper No.43.* Rome Italy Pp. 71-72.

FAO. 1999. *World Production of Animal Protein and Need for a New Approach.* F.A.O. Rome ACA Pp. 98 – 99.

Harris, F. 1995. Nutrient dynamics in the Kano close-settled zone. ILEIA Newsletter, December 1995,Volume 11 Number 4: 16–17.

IFSTC. 1984. International Food Security Treat Campaign. International Livestock Center for Africa. *Sustainable Product from Livestock in Sub Sahara Africa.* Programme Planning and Funding requirement. Pp. 93-97.

ILCA. 1998. *Annual Report of International Livestock.* Centre for Africa 1998, Pp. 43 – 45.

International Fund for Agricultural Development. 2005. Integrated Crop-Livestock Farming System, Burkina Faso. Rome: IFAD.

International Fund for Agricultural Development. 2008. Improving Crop-Livestock Productivity through Efficient Nutrient Management in Mixed Farming Systems of Semi-arid West Africa. http://www.ifad.org/lrkm/tags/384.htm (accessed 8 May 2008).

Jahake, Hans E. 1992. *Livestock Production Systems and LIvestock Development in Tropical Africa.* Kiel: Kieler Wiscenchaftsverlag Vauk. Pp. 72-74.

Kallah, M.S. and A.M. Adamu. 1988. The importance of animal feces as fertilizer. National Animal Production Research Institute, Ahmadu Bello University, Zaria, Nigeria. 9 pp.

Keftasa, D. and International Livestock Centre for Africa. 1988. Role of Crop Residues as Livestock Feed in Ethiopian Highlands. In Proceedings of the Third Workshop at the International Conference Centre. Arusha,

Kumar H., Singh J.N., Kadian V.S., Singh K.P., Saxena K.K. and Kumar H. 1994. Comparative productivity and economics of dairy enterprises under mixed farming systems. *Farming Systems* 10:36-44.

Singh K.P., Singh S.N., Kumar H., Kadian V.S. and Saxena K.K. 1993. Economic analysis of different farming systems followed on small and marginal land holdings in Haryana. Haryana *Journal of Agronomy* 9:122-125.

Patil B.R. and Udo H.M.J. 1997. The impact of crossbred cows in mixed farming systems in Gujarat, India: Milk production and feeding practices. *Asian-Australasian Journal of Animal Sciences* 10:253-259.

Van Keulen, H. and H. Schiere. 2004. Crop-Livestock Systems: Old Wine in New Bottles? In New Directions for a Diverse Planet. Proceedings of the 4th International Crop Science Congress, Brisbane, Australia, 26 September- October 2004. http://www.cropscience.org.au/icsc2004/symposia/2/1/211_vankeulenh.htm

Van Raay, J.G.T. and P.N. de Leeuw. 1971. The importance of crop residues as fodder: a resource analysis in Katsina Province, Nigeria. ABU, Zaria Samaru Research Bulletin 139.

Chapter 13

Strategies for Developing Sustainable Approaches and Models through Krishi Vigyan Kendra System

S. Prabhu Kumar and C.V. Sairam

Indian agriculture at present facing multi-faceted challenges both in domestic and international front. Climatic changes leading to erratic monsoon and other weather parameters, dwindling natural and farm resources, migrating rural population to urban areas and its impact on labour cost and its availability for farm operations, unfavourable and high degree of price risks are the major domestic challenges posed in the domestic front, while trade liberalization and globalization, demand for quality oriented agricultural products, increasing competitiveness in international markets are the major challenges put forth in the global front. At the same time the ever growing population at a fairly high rate demands higher growth rates in agriculture and its allied sectors.

In order to meet this demand, technology continues to play a crucial role. Researchers and policy makers opine that only 30 per cent of the agricultural technologies are being adopted by the farmers. Once the adoption rate is increased meeting the domestic demand would not be a major problem.

For increasing the rate of adoption of technology developed by National Agricultural Research System they need proper assessment and refinement before being disseminated on a larger scale. In this process the role of Krishi Vigyan Kendra

(KVK) is very crucial. This paper highlights the strategies for developing sustainable approaches and models through KVK system.

KVK System

The mandate of the KVK is

- Technology assessment, refinement and demonstration of technology/ products.

The activities of KVK include

- On-farm testing to identify the location specificity of agricultural technologies under various farming systems.
- Frontline demonstrations to establish its production potentials on the farmers' fields.
- Training of farmers to update their knowledge and skills in modern agricultural technologies, and training of extension personnel to orient them in the frontier areas of technology development.
- To work as resource and knowledge centre of agricultural technology for supporting initiatives of public, private and voluntary sector for improving the agricultural economy of the district.

In order to create awareness about improved technology, a large number of extension activities will be taken up. The seeds and planting materials produced by the KVKs will also be made available to the farmers.

The above mandate is carried out through prioritization and promotion of following thrust areas and identification of activities and technologies which has resulted in bringing out innovative extension methodologies and impact analysis through exploring cases of large scale adoption and assessing the anticipated and realized outcomes.

- Sustainable crop production through integrated nutrient management and organic farming strategies
- Integrated pest and disease management
- Introduction and popularization of improved varieties/hybrids of crops through technical and material back-up
- Development and promotion of crop diversification and alternate land use system
- Empowerment of women and youth through improved nutrition, income and reduction of drudgery
- Scientific management of dairy and small livestock
- Promotion of horticulture as a mechanism of crop diversification, augmenting family and national income
- Value addition, processing and market facilitation of household and commercial enterprises

- ☆ Soil, water conservation and watershed management for drought proofing and sustainable rainfed farming
- ☆ Small scale mechanization for saving time and reducing cost and drudgery
- ☆ Increasing income from fishery enterprises through production processing and marketing

These were mainly identified through linkage with development departments, discussion with research systems feedback, group meeting with farmers and through seminars/workshops. Thrust areas were promoted through target groups and by selecting technologies based on the need and interest among the farmers.

Most of the technologies popularized and promoted were location specific, result oriented, low cost, sustainable, eco-friendly, socially acceptable and were superior both technically and financially over the existing practices.

Promotion of the thrust areas by the KVKs reflected in terms of training (to farmers/farm women, rural youth, extension functionaries organised independently and collaborative mode), frontline demonstrations, on farm trials, extension activities, publications and production and distribution of quality seed material.

KVKs have evolved and practiced the following innovative approaches to reach the unreached and to enhance their extension efficiency with greater impacts.

- ☆ Group approach
- ☆ Seed village concept
- ☆ Veterinary link workers
- ☆ Future farmers
- ☆ Farmer participatory
- ☆ Information sharing
- ☆ Technology leaders
- ☆ Commodity quality promotion and partnership
- ☆ Institutional approach
- ☆ Technology spread
- ☆ Promotion of eco-friendly/resource conservation technologies
- ☆ Woman empowerment
- ☆ Production of input at site
- ☆ Information and communication technology
- ☆ Entrepreneurial development

KVKs plan and conduct activities with specific anticipation of consequences. KVKs could actually measure the extent of realization against the anticipated outcomes. In most cases, the extent of realization was more than the anticipated level. Large scale adoption of KVK technologies stands testimony to their utility in problem solving and profit maximization. Such cases of larger adoption are in the areas of alternate cropping systems

- ✰ Varietal spread
- ✰ Integrated crop production
- ✰ Integrated pest management
- ✰ Organic farming
- ✰ Farm mechanization
- ✰ Watershed development
- ✰ Fisheries
- ✰ Income generating activities and
- ✰ Group approach and marketing.

Present Strategies

The process being used by the KVKs in identifying and prioritizing the technologies are appropriate for the specific agro climatic and socio economic conditions of respective areas of their operation. The major thematic areas under which the technologies have been refined and popularized by KVKs include crop diversification and modification of existing cropping systems, introduction and popularization of improved crop varieties, integrated nutrient and pest management practices, alternate land use and dry land horticulture practices, post harvest value addition, and improved dairy and poultry management techniques.

The KVKs have significant contribution in dissemination of improved technologies through On-farm Trials (OFT) and Frontline Demonstrations (FLD). Assessing and refining technologies for their suitability to specific micro situations of each district and then popularizing them through well planned and executed frontline demonstrations have created tremendous potential for increasing the yield of major crops.

Among the various skills that were transferred to farmers and extension personnel, the major areas are improved production technologies for major field crops; horticulture crops including medicinal and aromatic crops; improved management techniques for dairy, goatry, poultry, fish and prawn culture; post harvest value addition of various agri-based rural products, and income generation activities for rural youth and farm women. The technologies and skills that have been transferred by KVKs through training are need based, appropriate to various clientele group and specific to their agro climatic and socio economic conditions.

The KVK adopted many innovative methods for transfer of technologies. Many of the extension strategies have taken into account the changes in social and economic conditions that are rapidly changing and are governed by regional requirements and ethnic conditions. The innovative extension methodology adopted by KVKs of the country could be broadly categorized into following approaches, i) Group approach, ii) Innovative farmer approach, iii) Electronic media and cyber extension, iv) Help line services, v) Traditional and folk media, vi) Interactive and exposure visits, vii) Cluster area approach, viii) Own or hire your inputs and services, ix) Farmers field school, and x) Para veterinary technicians.

Table 13.1

Sl.No.	Existing Approach	Alternate Approach	Evolved Strategy	Expected Benefit
1.	Problem identificationand prioritization basedon individual basis	KVKs needs to group themselves on agro-climatic basis based on NARP/Planning Commission classification and needs to indentify the common problems and prioritize them using appropriate PRA techniques.	Common technological interventions for a specific agro-climatic zone	Impact of the technology can be assessed on a larger scale
2.	Technology interventions common for various operational areas in the district	Differential means of technology interventions in the district based on the existing productivity level	Technology intervention based on productivity level. For example level and nature of intervention will vary according between high productive and low productive zone	Lesser productive zone in the district may be specifically identified and intervened in order to realize real benefit from the technology
3.	Group approach in production	Group approach in production, processing and marketing	Aiming to change farmers from producers to Agribusiness Entrepreneurs	The role of farmer is changed as businessman
4.	KVKs and farmers does not play their essential role in Supply Chain Mechanism	Effective linkage between producers and consumers	KVKs and farmers act on a Supply Chain Mechanism mode	
5.	Lesser emphasis on quality consciousness production	Additional emphasis on quality based production	Increasing quality consciousness among the farmers especially in group approach	Lead to export oriented production
6.	No role in encouraging investment back to agriculture and its allied activities	Encouraging farmers to invest back in agriculture and its allied sectors for sustainable development of rural environment	Aggregate Investment in agriculture and its allied sectors increases	Sustainable growth and development of district agriculture and its allied sectors
7.	Promotion of cropping system	Promotion of animal husbandry system (rearing of group of animals in the same farm on a medium or larger scale)	Integrated Animal Husbandry Based Farming System	Shifting the prime source of income from crops to animals. This system may be successful especially in rainfed areas

Contd.

Table 13.1–*Contd.*

Sl.No.	*Existing Approach*	*Alternate Approach*	*Evolved Strategy*	*Expected Benefit*
8.	Lesser emphasis on employment generation in technologies	More emphasis on employment generation on various technologies in agriculture and its allied sectors	Convergence of KVKs with NREGP and SGSY. KVKs can provide technology backstopping to the employment guarantee schemes in their district	Prevent migration of labour force from rural to urban areas
9.	KVKs does not have adequate data on district agricultural and its allied sectors	KVKs need to act as database centre on vital statistics of district agriculture and its allied sectors	Database Management as a mandate of KVK	KVKs can scientifically advise and involve themselves in planning process of agriculture and its allied sectors
10.	Moderate level of Information and Communication Technology	Strengthening the level of Information and Communication Technology in KVKs through e-linkage	Emphasis on higher degree of usage of ICT	Faster and apt way of technology dissemination

Major Areas of Impact

Participatory Impact Monitoring and Assessment (PIMA): The nature of impacts to be studied, indicators to be used, and development of data collection tools, methods of data generation and the interpretation. True participatory approaches were employed to develop various skills among participating officials of the KVK. Impact of this concept has already been seen as many of the KVKs have taken up the impact assessment of their activities.

Based on the methods used by KVKs in assessment of mandated activities, KVKs in the country have made considerable impact in several areas. Some of such areas, where impact is visible and significant are crop production, seed production/ replacement, integrated pest management, natural resource management, livestock production and management, and fisheries, information and communication technology, entrepreneurship development, and innovative extension approaches. Some cases of large scale adoption of technologies disseminated by the KVKs are furnished below:

- Improved crop management strategies
- Crop diversification and alternate land use practices
- Income generating activities
- Development of seed villages
- Popularization of millets
- Adoption of horticultural crops
- Adoption of dry land technologies

Alternative Strategies for Sustainable Development

Changing domestic and global agricultural scenario demands alternative strategies and plan of action for KVK system without much alteration of the existing mandate and approaches. For this the following strategies would help the KVKs to formulate and implement sustainable approaches and models to achieve their mandate to a satisfactory level.

Conclusion

Global and Indian agriculture is getting transformed as a means of livelihood to means of business. In this process, farmers are transformed as mere producers to agribusiness men. Though technologies developed by National Agricultural Research System would help them for this desired transformation, KVKs have to necessarily adopt certain strategies for application of these technologies by the farmers in their district. Adoption of common technological intervention based on agro-climatic zones, technology intervention based on existing productivity level, inculcating the concept of agri-business management among the farmers, encouraging investment in agriculture and more emphasis on ICT are some of the major strategies need to be evolved and implemented by the KVKs in the XI Plan period.

Chapter 14

Knowledge Networking in Agriculture and Outreach Challenges

A. Sarkar, A.K. Singh and P.P. Pal

Background

Present Scenario

Achievement of self reliance in food grains is the greatest success of agricultural extension with the solid support of research system. But the challenge is not the same today. In spite of such spectacular achievements, India still stands at 127th position in the globe in terms of human development index. Social exclusion in terms of low agricultural growth, concentration of poverty among the weaker sections, inequalities and inter as well as intra regional disparities has been a focused concern now. The country entered into new millennium with around 260 million people not having income to access a consumption basket that defines the poverty line. Of these, 75 per cent were in the rural areas. India is the home to 22 per cent of the world's poor. Though the X plan document of planning commission perceived that reduction of poverty in the country was vital for the attainment of international goal, the coping strategies adopted to beneficiary targeted Poverty Alleviation Programmes (PAPs), however, could not make much dent due to the top-down and over – determined approach (Moni 2004). Alongside, and as a fall out of trade liberalization in post WTO scenario, competitiveness in the agricultural pursuits has increased substantially to create both crisis and opportunities for the Indian farming gentry. While the resource poor and disadvantaged rural masses have started facing the heats of liberalization in the form of gradual reduction in subsidies over various agro-inputs and un-preparedness on their part to match with the emerging knowledge and skill requirements to effectively counterbalance the market led production scenario, new vistas have also opened up, based upon agricultural produces and products, for the increased consumer oriented economic activities within and outside the country. At this juncture, Indian agriculture also needs a productivity and quality revolution *i.e.* a Total Quality Management (TQM) and a Total Factor Productivity (TFP) which can be brought out through the much-needed agricultural reforms.

How to Meet the WTO Challenges

- Empowerment of resource poor farmers with suitable technical knowledge and technical skill for cultivation of improved crops. This is possible only with effective and meaningful combination of different extension teaching methods.
- Information support and choice facilitation for taking production decision, marketing of outputs, business support, market shocks etc.
- Efficient extension workers to impart practical and useful skills through effective and timely communications.
- Creation of better opportunity for the farmers and other primary stake holders for outside field exposure to understand various techniques and field operations carried out by other farmers in matters of production technology, diversification,
- On and off farm value addition, group-led activities, coping strategies to bail out from perceived market shocks etc.

Extension: A Key Building Block

As we understand, the primary goal of agricultural extension is to assist farming families in adopting their production and marketing strategies to rapidly changing social, political and economic conditions, so that they can, in the long term, shape their lives and livelihoods according to their personal preferences and those of the community. Agriculture is different across various parts of the country and, hence, the *one-size-fits-all model (i.e. Green Revolution Model)* of agriculture is inappropriate and cannot reflect the environmental, social and cultural diversity that exist around the country. *Extension strategy for rain fed, fragile and vulnerable small production systems that often operates under complex diverse and risk prone environment is different than irrigated, well endowed and resource rich farms.* In her bid to pursue with the goal to help overcoming twin challenges of livelihood promotion and sustained economic growth through technology intervention, business development and marketing, the importance of extension services can not be over emphasized. Under this challenged scenario, the outreach delivery system needs to be efficient enough to solve both the *first mile* and *last mile* problems through cost effective, easily deployable and maintainable information mediation.

Future Challenges for Extension and its Managers

In the emerging scenario, the task of extension has become to be more challenging which demands a *system of market-led extension* with specific focus on diversification, post harvest management, export orientation, market intelligence and consumer orientation. However, the most important challenge for the future extension system and its managers would be the *Management of Knowledge.* In the coming years, the success of a farmer is going to be primarily dependant upon his level of knowledge. Agricultural extension programmes need to be oriented towards enhancing the decision making abilities of rural people, especially farmers, by expanding their exposure to effective organization and management skills in a *knowledge-intensive manner* and not just new agricultural production technologies alone.

Key Tasks Ahead

- ☆ Reorienting the standard of farmers' socio-economic level
- ☆ Reorienting the extension system to effectively address post WTO compulsion
- ☆ Production and market oriented communication
- ☆ Strengthening of market information sources including price, storage and transportation
- ☆ Linking with weather forecast information
- ☆ Concentrating on quality and value added production procedure
- ☆ Dissemination of post harvest technology
- ☆ Educating the farmers about sorting, grading and other market procedures

Knowledge Connectivity: A Means for Rural Growth

Today, India is both a developed as well as a developing nation at the same time. Extremely advanced forms of agriculture and industry co-exist with other forms of yester years. Bridging this gap is the ultimate development challenge of 21st century. To accelerate the march towards rural rejuvenation, a wholesome and forward looking scheme needs to be devised. In order to achieve our national goal to be a *Knowledge Society*, we can, therefore, launch a learning revolution in rural India through the establishment of *Village Knowledge Hubs* involving the integrated use of both the conventional as well as upcoming electronic media like Intra and Internet, Information Kiosk, Cable T.V., Radio (including HAM, Community Radio), Mobile Telephones, Vernacular Press and other Print Media.

Farmers' Knowledge Society – Why?

- ☆ Knowledge empowerment and global access
- ☆ Social Mobilization of the Community
- ☆ Empowering rural people in decision making
- ☆ Increased Food Security through knowledge based agriculture
- ☆ Increased on-farm and non-farm employment
- ☆ Increased capital flow and higher economic security

ICT in Developing Village Knowledge Hubs

ICTs provide us a new opportunity to build up a confident, skilled and participatory knowledge community that includes women also. Introduction of ICT, as has already made inroads into our agricultural scenario, would be immensely helpful to collect, store and share information between people through using multiple devices and multiple media so as to bridge the perceived knowledge gaps at the rural nexus. In spite of high possibilities, ICT did not percolate down evenly to all sections of the population. Only a fortunate few is familiar with the potentials these technologies hold. The skill to realize the potential is also limited- those who have are naturally better placed to prosper in this competitive world. And this uneven diffusion of the awareness, skills, infrastructure etc.among the potential users has given rise to ***digital divide*** (Lahiri 2004). To effectively cater such divide, we can offer a range of information inputs and services through effective use of ICTs.

Agro-Advisory Services

- ☆ Disease forecast service
- ☆ Online advisory and consultancy
- ☆ Computer to Mobile/SMS Broadcast Service
- ☆ *Krishi Gyan*

Agri-Business Management

- ☆ Input Sector
- ☆ Farm Production Sector
- ☆ Processing Sector
- ☆ Marketing and Distribution Sector
- ☆ Information Support Sector

Weather-based Pest and Disease Forecasting

- ☆ Pest and Disease Database history
- ☆ Short and Medium Range Decision Support on Plant Protection
- ☆ Annual Crop Planning

Weather Information for Decision Support on Crop Management

- ☆ Weekly report on crop-weather conditions
- ☆ Soil, Water and Climate outlook, Radiation
- ☆ Production potential
- ☆ Seasonality and weather aberrations

Method of Soil Fertility Evaluation

Quality Management Requirement

Quality Reporting Procedures

Certification Procedures

Product Information Guide

Sources of Laboratory Testing Facilities

Answer to Frequently Aske Questions (FAQs)

Interactive Voice Response Service on Local Weather Forecast and Market Information

Online Consultation with Farmers

Net School for Farmers

Key Areas for Action and Intervention (A+ 8C Framework)

A for AWARENESS and ADVOCACY

- ☆ Aware of the implications of information and knowledge revolution
- ☆ Visions and strategies to adopt effective legal frameworks
- ☆ Policies and action plans at National, Regional, State and Local levels

First C for CONNECTIVITY

- ☆ Building up of small scale bottom up Rural Information System for extending last mile connectivity through harnessing around 7,00,000 km of buried fiber optics cable as well as wireless technology
- ☆ Physical and material access for both the traditional and electronic media including broadcasting and telecasting.
- ☆ Integration of Community Radio Stations with Local Systems
- ☆ Application of HAM Radio and Broad version of existing wireless access system CorDECT WLL Technology as per fitment of locations to ensure faster data transfer rates.

Second C for CAPACITY DEVELOPMENT

- ☆ ICTs require both institutional and human capacities.
- ☆ Over and above the proposed support under national level capacity building programme for community knowledge workers within e-Governance Action Plan (NeGP), capacity building among community members is also suggested to be included for ensuring future effectiveness of Village Knowledge Hubs.
- ☆ Developing focused linkages and partnership with local institutions, NGOs and Community Based Organizations (CBOs) for identification of master trainers and building up their capacities.

Third C for CONTENT

- ☆ At present, English dominates the internet. Some 80 per cent of all websites are in English compared to less than 0.0001 per cent of internet content today in Arabic (Das, 2006).

- Creation of digital and other content, duly keeping in mind the linguistic and cultural diversity exists in the country.
- Provisioning of "knowledge broker" and solutions websites.
- *As considerable section of rural India is not computer literate, multi-media, audio and video interfaces and technologies that work on a touch screen system need to be initiated, to make the content user friendly.*
- As Rural communities are innately knowledge intensive and practiced as well as inculcated a culture of knowledge since ages, for effective blending of local knowledge with those of cutting-edge scientific ones, hence, rural communities should be effectively linked with content template development process for **Village Knowledge Hubs**.

Fourth C for CREATIVITY

- PROVISIONING FOR Solar dynamo-driven computers to befit rural areas that are void of electricity
- Innovativeness as per local resource base to protect the knowledge hubs against sand, humidity or other external influences
- Opportunities for voice based software programmes targeting more on illiterate persons
- Digital innovation approaches
- Digital community tele-centres

Fifth C for COMMUNICATIONS AND NETWORKING

- Need to widen participation and involvement in decision making at all levels to
  - foster societal empowerment,
  - governance,
  - transparency and accountability,
  - stimulate networking – nationally, regionally and globally

Sixth C for COORDINATION

- As the village knowledge hubs have been conceived to be community centric and local demand driven, development of effective coordination and linkage between the local and district level institutions – PRI, KVK, ATMA, Academic and Financial Institutions, SHGs, NGOs, CBOs and even existing PCO Operators seems essential.

Seventh C for COALITION AND CONVERGENCE

- Challenges are imposing and so is the necessity of effective synergy through partnership and convergence.
- Only public sector investment through *NeGP/Bharat Nirman Programme* won't be enough for such ICT 4D activities. Consortia based initiatives with participation from Government organizations, private sector, academic

institutions, civil society organizations, CBOs and even international agencies are essentially required.

Eighth C for CASH, CARE AND MANAGEMENT

- ✰ The scope is immense and the experiences gained from pilot running of different projects in the country are also rich. There were the initiatives from the public sector, NGOs, civil society organizations, ICAR, KVK and even from the corporate giants like ITC etc. The need is now to have a network of consortia across the country with defined apex bodies for national, regional, state and district levels with specific terms of reference.
- ✰ The amount of possible fund flow from different sources like *Bharat Nirman Programme*, RIDF scheme of NABARD, state fund for e-governance, programme based farming from international authority like DFID etc. should be ensured through coordinated planning.
- ✰ Mechanism to be derived for gradually grooming the village knowledge hubs to be self sustaining service providing centres and, hence, those be allowed to impose fees against different multitude of services.

The Way Ahead

The learning experiences on ICT under Indian environment have proved beyond doubt that even the poor rural communities can be effectively facilitated with information to, in turn, make impact on their livelihoods. It, however, needs to be recognized that like any other induced change, change through ICT is also a slow process. It is the joint responsibility of all the stakeholders now to facilitate the ongoing ICT revolution so as to touch the lives of the rural poor so that the job of reaching the unreached can effectively be addressed in the coming years. We are to keep in mind that it is only when the ICT mediated rural knowledge hubs will continue to perform in a scalable manner, even after withdrawal of donor funding, we may only then be able to place this immensely important extension means on a sustainable platform.

The End Note

The country has made commendable stride in the ICT based knowledge initiative. The learning experiences from multimedia community centres or telecentres through pilot projects have ushered a major paradigm shift away from individual connectivity. What we are to do now is to groove out of the pilot syndrome and develop a national network of Village Knowledge Hubs to foster India achieving its cherished goal to become a knowledge society. And the role of extension system here is crucial. Building mass awareness, developing local level linkages, facilitating usable and demand driven content development, capacity building of the stakeholders for managing the Village Knowledge Hubs and empowering the rural managers for rendering sustainable and scalable services are the tasks ahead for extension managers.

References

1. Lahiri, A. (2004). Community Information Services – Policy and Planning Issues. Paper presented in the National Seminar on Information and Communication

Technologies for Agriculture and Rural Development held at NAARM, Hyderabad. pp 73 – 81.

2. Moni, M. (2004). Agricultural Resources Information System (AgRIS): An e-Government programme for fostering agricultural growth, poverty reduction and sustainable resource use in India – a step towards establishing a location – specific e-Government model for the poor. Paper presented at the Regional Workshop for implementing e-Government at U.N. Conference centre, Bangkok (Thailand), organized by Asian Development Bank Institute (Japan) and UN/ ESCAP, Bangkok.
3. Das, P (2006). Developing a Hub of Knowledge in KVK through Information Communication Technology. Proceedings of 2nd National Conference on KVKs held at Acharya N.G. Ranga Agricultural University, Rajendranagar, Hyderabad.

Chapter 15

Partnership in Agricultural Extension: Needed Paradigm Shift

Baldeo Singh

In the post green revolution era of agriculture when liberalization and globalization are the buzzwords the demand for agricultural technology are changing and diversifying. To meet the food demands of the growing population and to compete in the international scenario, technologies are needed (1) to break the stagnation of yield growth in intensive irrigated agriculture; (2) to harness the production potential in dry land farming which accounts for majority of India's cultivable land; (3) to make optimum use of natural resources for sustainable production; and (4) to make investment in agri-preneurship development.

Due to economic liberalization and signing up of the WTO agreement, the market place has become uncertain and volatile. There is desperation and confusion among the farming community as to what to grow and where to get the information to make best use of their limited resources. There are also lots of private players cropping up in the field of extension providing various kinds of services. We are also experiencing the unfolding of the information technology revolution. The advancement of the technology on the information and communication front is posing its own challenge. The rapid development in agriculture has resulted in increased importance of extension services during last few years. The role of extension in transferring technologies has been fairly acknowledged. The effective performance of extension system is now a real challenge for following reasons:

1. *Information now has real, measurable value:* Private-sector information providers such as crop advisors, business consultants, and private education and training providers are cropping up everywhere. The

attraction is the revenue generated by refining information to meet specific audience needs.

2. *Public extension services are no more sole-source of information:* Extension has spent decades as a sole-source provider in the information and outreach market. Technology has created the opportunity to provide and promote access that wasn't realistic before. Because of this, the traditional educational market is now more easily accessible for competitors - agribusiness and commercial crop market advisors. Technology combined with open access to the rapidly expanding knowledge marketplace entices private information providers to step up the competition for clients (farmers) who previously were dependent on public sector extension as a sole-source education provider.
3. *Essential shift from "provider mentality" to "user mentality":* A "provider mentality" focuses on the process of distribution, meaning it is driven by what we want to distribute. A "user mentality" focuses on *access* and the customer, meaning we now need to anticipate customer needs and provide them *access* to our knowledge base. Therefore, there is need for traditional extension system to shift from distribution to access.
4. *Required shift from broadcasting to narrow casting:* The satellite T.V. makes available specialized channels for the access to all sorts of information including music and sports channels. Similarly, database is being created for almost every subject matter area be it be pesticides, chemicals or any thing else. People want to choose where they get specific information and not lock themselves into choices made by one provider.
5. *Instance Performance: The* market place demands customized, immediate response to problems and questions. Therefore, the challenge is to synthesize the data and make effective need based prompt decisions.
6. *Demand driven and customized information:* In this demanding scenario, the challenges to extension system are many and varied. The extension must satisfy the clients with location-specific and need-based information.

Extension educator needs to:

- ☆ Learn with clientele rather than being their teacher
- ☆ Overcome information overload by getting information to the client when they need it
- ☆ Adapt the information to local needs and values
- ☆ Help both the economically productive sector that has access to technology and those without resources that want to acquire and use new technology
- ☆ Make sure the extension educator is a knowledgeable and trusted consultant

In the present scenario, the role of extension is much wider and it needs to change its focus from safeguarding national food security to commercialization and export crop production; to equip themselves with traditional skills understand the clients and the local situation; train farmers in management skills and decision making skills; help the rural people develop leadership and organizational skills enabling them to organize better and participate fully in the development programmes; and develop their skills in marketing, selling, negotiating, delivering and using the knowledge of latest communication technology

Public and Private Extension

The demand for agricultural information by farmers is not uniform across the producers and the regions. After independence, the country initiated a number of programmes and created several institutions to provide extension and other services to the farmers. The information on agriculture is being provided by various public and private agencies. The public sector extension has the larger coverage, while private and other extension providers have limited coverage, rather supplementary role to a greater extent. Over the years number of organizations performing extension functions has been increasing. The performance of public sector extension is under scrutiny for quite some time. The questions are being raised on its capabilities to deliver goods in rapidly changing environment and this resulted into increasing number private players in the extension field. From 1980s, the need for involving NGOs, private sector and farmers associations in sharing, augmenting and supplementing public sector extension is increasingly recognized.

The brief review of the extension performance and a sketch of the present changing scenario and its challenges in general and to extension in particular indicate a very gloomy picture and a serious gap towards competitiveness and quality performance. Usually extension has been used to meet expanding demands for food and to cope with declining availability of various resources. Its educational role in terms of problem diagnosis, effective transfer of technological messages though educational aids, feed back of farmers' response to technology and development of effective linkages with different agencies to solve problems of farmers; particularly in public extension is almost missing. The private extension systems are maintaining good contact with their clients and working in interpersonal and educational mode to some extent. This is mainly due to their targeted approach and limited area of jurisdiction. It is true that private extension cannot replace public extension. Public extension has to remain the main player in the field of extension. Many debates are going as how to make public extension system socially accountable and to what extent and in what area the private extension system is allowed to supplement the extension. What should be our strategy to cope with these challenges and how to bring about the needed changes in the extension system to make it more responsive and client oriented are crucial points need to be addressed appropriately. The argument in this connection is as under:

1. Fund crunch: In the era of disinvestments and privatization, the government machinery is not able to bear the burden of expenditure on public extension services. To overcome this fund crunch increasing privatization is the answer.
2. Efficiency and effectiveness: Reasons for efficiency and effectiveness may be many but there is a feeling that privately runs organizations are more

efficient and effective resulting in cost reduction and increased profits and optimum use of the resources.

3. Demand for specific information: The private information provider works on the specific areas and can meet the demand for the specific information more efficiently than public extension system, which is nearly paralysed.
4. Paid services used more effectively: The past experience has confirmed that the paid services are more effectively used than the services provided free of cost. The clients always regard such services as reliable. There are areas where paid services are working well.
5. Increasing Transformation: The increasing commercialization of agriculture and innovative development in information technology have provided new sets of problems, opportunities and challenges, which require client friendly, market friendly and export/trade friendly interventions. A market is developing for skilled and specific agricultural advice. The private extension providers have comparative advantages to this situation and can play the desired role in this market.

Now the question is can we submit to such a radical view and privatize our extension completely. If not, what is the way out? The way out is the careful integration of public and private extension services according to various needs, so that the two can complement and supplement each other.

Thrust Areas of XI Five Year Plan

Agriculture and rural developments are two sides of a coin in sustainable development at gross root level. Acoordingly, in the thrust areas and programmes of agricultural extension in the XI five-year plan, this aspect of integration of private sector extension services with present public system has been stressed on following ways:

- ☆ *Promotion of Agribusiness through development of Agribusiness Center:* Under the new programme initiatives, there is proposal for promoting agribusiness through development of Agribusiness Centers. The objective cannot be achieved without the partnership with the private players.
- ☆ *Establishment of Agro-Clinics:* The demand for the specific farm information is on the rise on one hand and on the other hand, a large number of agricultural graduates are in dire need of employment and they are seeking job opportunities outside. So, they can be trained to open these information shops. These shops can render advisory services on both crops and animal husbandry and act like single window for farm products, input information, inputs sale etc. at the block and district levels. Private agencies can play active role in successful implementation of this project.
- ☆ *Consolidation and strengthening of KVKs:* At present, there are above 600 KVKs sanctioned, a reasonable numbers of them are run by NGOs. Now the question is how to bring about this integration? The process requires thorough understanding of the strengths and weaknesses of both the systems *i.e.* Government Organizations and Privately run institutes.

- ☆ *Contact farming:* There are many MNCs coming up in big way for contract farming *e.g.* Pepsi in Punjab and ITC in Andhra Pradesh. The tomato cultivators of Punjab reported that the new technology introduced by the MNC and due to effective extension and market provision provided by MNCs contract farming led to cost cutting and higher yields consequently higher income both for the farmers as well as labourers.
- ☆ *Higher marketing efficiency:* Experiences have shown that it is impossible to attain sustainable production isolated from marketing situation. There are number of such examples in which farmers become the victims of such extension approaches that emphasized to produce more without proper attempt to organize marketing. Under such circumstances as the farmers incurred heavy losses, private agencies can help by providing:
  1. Information about market opportunities, price and marketing technology transfer.
  2. Forward and backward linkages for increasing marketing access to farmers.
  3. Post harvest technology to avoid distress sales including processing, grading, value addition, storage, packaging, transportation, distribution, product standardization, and reducing number of intermediaries.

The MNCs have the advantage of strong capital foundation and they provide enormous employment potential. Some of these MNCs are also having big R&D setup.

Implications

Our country's development stage has a bearing not only on the prevailing levels of technology but also on the demand for agricultural produce and difficulty of providing extension. Most of the technologies for basic food crops will always remain a public good and hence should be provided by the public sector. The implication pointed here is that narrow goals of economic efficiency cannot be replaced by a broader aim of enhancing general welfare. While advocating institutional pluralism, a broad conceptual vision of extension is needed, so that any systematic link between two sides is not neglected because of increased complexity of bringing all sides together.

Government funding in India should continue to focus on public goods activities such as technology transfer, education and human resource development, while private extension should focus on specific needs of special client groups. Our government can take on a key role in developing strategies for the evolution of an extension system, which takes into account subsidiarity and complemtarity among actors. It is important to devise a public extension policy because the extent to which other bodies will be willing to provide extension services is partially determined by government actions. Government should provide an enabling environment for non government provision of extension services in which the players come together share experiences and plan future and rural people's needs are more efficiently served on sustainable basis.

Opportunities for successful integration of the efforts of public sector, private sector and farmers groups are emerging in some areas. The need is to develop location based strategies and varied combinations of financing and providing extension services in different situations. At least block level and below, there should be close integration among public agencies, private agencies and farmers associations. There is a need to have adequate structural reforms to commercialize agriculture at this level for meeting the pressing needs of food demand, job requirement, open market for export/trade avenues and skill development in human resources under one umbrella with minimum political interference and maximum social accountability. The private agencies should be given full opportunity to play its role at this level. This is essential for economic and social reasons.

IARI Initiative 'Village-Based Integrated Models for Market-Led Agriculture'

The extension initiative of IARI is to develop 'Village based integrated models for market-led agriculture' in different parts of the country by involving institutions. The programme aims at creating location-specific models to utilize properly prime land and rural resources for agriculture; realize the untapped production potential by bridging yield gaps; ensure household food, income and employment security; develop system approach with all attention to the links in production to food value chain; develop local capacity for participation, leadership, group action and entrepreneurship; sensitize, upgrade and strengthen the extension support and linkages; strengthen input-output chain from farm to market and sustain retail boom and promote quick spread of technologies through farmer-to-farmer extension in wider areas.

The IARI has established project field sites in north, south, east and west parts of the country by involving institutions to carry out TOT interventions in participatory mode in the fields of diversified/commercial farming. The attempt is to understand the problems faced by the farmers related to their livelihoods and adoption and diffusion of farm enterprises, and address them to strengthen the technologies/enterprises movements through farmer-to-farmer extension. The local partners being involved include development departments, SAUs/ICAR institutes, NGOs or corporate sector. Some selected nearby sites are being connected with IARI for information communication and knowledge sharing. The project is being carried out in phased manner in concentric mode in collaboration with selected SAUs/ICAR institutes across the country.

1. Peri-urban agriculture (up to 75 Kms. from IARI)-Three (3) sites in NCR of Delhi + Gurgaon (Haryana)
2. Integrated agriculture (above 75 to 500 Kms. from IARI)- Eight (8) sites (Sonepat in Haryana, Western districts and Mathura in UP, Patiala in Punjab, Churu, Jhunjhunu and Bharatpur in Rajasthan)

3. Integrated farming systems (beyond 500 Kms. from IARI) - Nine (9) sites (Gulbarga and Bangalore in Karnataka, Parthian and Rahuri in Maharashtra, Varanasi and Mirzapur in UP, Ranchi in Jharkhand, Udaipur in Rajasthan, Navsari in Gujarat)

A cluster of 2-4 villages/a village *Panchayat* has been selected at each project site. The innovative farmers in different project locations are also being linked and utilized for technology transfer. A mutual agreement for collaborative efforts between the IARI and SAUs/ICAR Institutes has been made. The SAUs/ICAR institutes involved are MPUAT, Udaipur (Rajasthan), MPKVV Rahuri (Maharashtra), MAU Parbhani (Maharashtra), UAS Dharwad and UAS Bangalore (Karnataka), BAU, Ranchi (Jharkhand) and BHU (UP), IIVR, Varanasi and CIRG, Mathura (UP), NAU, Navasari (Gujarat) and NRC Rapeseed-Mustard, Bharatpur (Rajasthan) from Kharif 2007. The programme encompasses village-based natural resources management and lays emphasis on quality seeds production, protected horticulture, diversification with high value crops/vegetables/flowers, processing, value addition and marketing, capacity building for entrepreneurship development by integrating indigenous knowledge and practices. The exchange of technologies among partners for trial and adoption in suitable locations and avoiding duplication of research efforts are also the concern of IARI initiative. It involves a farming systems approach with following emphases:

1. The concept of Seed Bank by producing enough good quality seeds and planting materials of important crop/vegetable varieties and hybrids through appropriate training and education to meet requirements of village cluster and nearby areas.
2. The concept of Water Bank by harvesting rain water and judiciously using it by adopting efficient water management techniques and choosing the crops and varieties according to available water quantity.
3. Development of Fodder Bank by introducing quality fodders and their conservation techniques using Feed Block Making Machine through farmers' association/SHG//*Panchayat* for successful rearing of animals for dairy, meat, wool, draught purposes.
4. Development of Energy Grid/Energy Bank by harnessing renewable energy from different sources such as solar, wind, water, biomass, bio-diesel etc. to meet major demand of energy.
5. Development of Manure Bank by utilizing cattle dung and other bio-mass for improving soil health and physical properties, providing renewable energy and reducing pollution.
6. The concept of Knowledge Bank by developing knowledge society through information support in production, processing, value addition and marketing.
7. Development of Villagers Cooperative Bank completely owned and managed by the rural people with appropriate directions and risk mitigation fund for helping the resource poor under unforeseen adverse situations.

The IARI extension is to create 'Village-based Models of Market-led Agriculture' in various locations so as to improve livelihood security and facilitate adoption and diffusion of technologies through farmer-to-farmer extension. The model envisages building up knowledge base of bio-diversity experiments of agro- enterprises; promoting rural knowledge systems; empowering people and their institutions for sustainable management of rural natural resources; developing local capacity for effective implementation of agro-enterprises through field demonstrations/research and information networking and establishing partnerships among various institutions involved in agro-enterprises development.

Chapter 16

An Innovative Approach to Field-Extension: Paliganj Experience

B.P. Sinha

Field extension may mean many things to many people, but what I mean by this is the production of desirable change in farmers'(human) behavior which is reflected in their fields (visibility of change). For this to happen effectively and efficiently, our communication must not be 'prescriptive' (which is so common), but explanatory, comprehensive and scientific in nature with explicit or intelligible logic and reasoning.

How was this approach in communication achieved would be explained with live examples drawn from the field experiences. Prescriptive approach is quite common with which almost everybody is familiar. In this approach, the client is given the package of practices with the advice to adopt them in the same sequence/order without raising any question with assurance given that if this package is adopted, the crop yield will substantially increase, profitability will go up and economic condition will improve (or some times economic condition will leap frog.) The prescriptive approach derives its energy or drive from the assumptions that the farmers are ignorant of their development needs and potentials due to their being uneducated, tradition oriented, incapable of understanding science and its logics and they are accustomed to have blind faiths. Thus they are told 'to do this and don't question this'; 'you can't understand beyond this and so follow this blindly'. This is a serious underestimation of the farmers and their abilities and thus this approach sometimes becomes dangerously counterproductive. I would like to give an example of this in the following anecdote.

> Once we visited a village in Gurgaon district of Haryana. Having heard that scientists from IARI had come, a good number of farmers gathered quickly and started making strong request that you should visit our village regularly and help us in solving our farm problems, I suggested them that 'you have an agriculture university at Hissar and its K.V.K. at Gurgaon itself. Why don't you visit them to get your farm problems solved'. On this one of the farmers got quite excited and almost shouted, "Sahab Unki baten mat karo we to ghane bebkoof hain" (Don't talk of them, they are great fools). I was quite stunned and asked how do you say this? In his excitement, the man ran to a nearby farm and returned in a few minutes with a white grub in his hand. He said (showing the white grub), "this is a great menace to our crops. I once went to them with this and asked them the remedy, they suggested 'light trap' as its control measure. Now you tell me can this fly? If not then how will it go to the lamp flame and die"? I got the clue. The scientist simply prescribed the control measure without talking about the life cycle of the insect and without explaining that the moths would get attracted to the light flame and would get dropped in the water mixed with kerosene oil kept below the flame. This would reduce their number resulting in reduced number of eggs laid and subsequently reduced number of white grub and hence reduction in damage to the crops. See the danger of prescriptive communication and implied effectiveness of explanatory and comprehensive commutation.

In our field extension work we resorted to the explanatory or comprehensive approach in communicating technology to them. For this, we followed the following steps –

(1) Clear identification of issue to be addressed to: - One of the issues identified in our project was 'very poor yields of crops despite lands being very fertile with assured irrigation facilities through a network of Sone canal system in Paligunj area of Patna district. Take for example the most major crop - rice which occupies over 96 per cent of cultivable land, had an average yield of 20.13 quintal/hectare as against the potential of over 50 quintal/hectare.

(2) Analysis of the existing package of practices to identify the practices which are counter productive, or which inhibits the production potential of the plants: it is quite obvious that for such a low yield the cultivation practices in vogue must be responsible: This analysis clearly brought to light a number of practices being followed by the farmers which were counterproductive or production inhibitive. Some such practices, for example, are

(a) Seed rate – In nursery 25 to 30 kg of paddy seeds used to be sown in 1kattha (1/32th acre) of land for raising seedling to transplant 1 acre of land. Obviously the plant population in the nursery are exorbitantly high. Seedlings are too close to each - other. The lands are available to each plant for nutrient up take was too small and too competitive, resulting in grossly inadequate nutrients supply to them. Hence the seedlings were unhealthy, slow growing and fragile. The consequence was that the seedlings used to take 50 to 60 days to become transplantable, they lost their vigour, many got broken while uprooting for transplantation and hence after transplanting, they remained poor in growth, tillering and earhead bearing giving lower yield per unit of land.

(b) No. of Plants Per hill and overall plant population – Seeding being very thin, weak and some broken during uprooting, 10 to 15 plants per hill used to be transplanted. This practice became counterproductive in variety of ways –

(i) The seedlings were too competitive in to deriving nutrition from soil.

(ii) Because of large number of seedling per hill planting distances were kept wider to cover the entire field resulting in less dense plants population.

(iii) Because the seedlings were weak and tall (50 to 60 days old), they used to be placed deeper in the mud so that they could remain standing, Hence first node of the plant (considered the best for producing healthy tillers) were inside the soil and tillering used to take place from subsequent nodes 15 to 20 days after planting. Consequently the tillers used to hear small ear heads and the later ones remain without bearing any ear head.

(iv) Because of poor health the seedlings used to fade way and top portion try up during the period they establish their roots in the soil after transplantation and start taking nutrition from the soil (about 6-7 days after transplantation). After they establish their roots in the soil, they produce tillers; hence initial tillers become the main plants after drying up of the tops. This reduces the life span of the plants since maturity dates remain unaltered resulting in poor productivity.

There were many other counterproductive practices like home grown poor quality seeds with high mixture and low vigour, improper selection of variety, late sowing, defective fertilisation of land, poor weed and disease pest control etc.

(3) Finding out substitutes to replace the counterproductive practices – After identification of counterproductive exiting cultivation practices, the substitutive practices were identified from the recommended package of practices with full scientific rationale and then a communication package was developed.

(4) Communication strategy – The package of practices selected to constitute the local practices were not communicated to them in a prescriptive manner. An approach of participative communication was adopted. For this, training camps were organized in selected villages in which group interactive

process was started with discussion on the local practices being followed by them, and how they were counterproductive taking individual processes one by one. The scientific reasons behind the counterproductive impact on plant growth processes were discussed in simple language with similes or resembling processes drawn from day to day life. For example, comparing weak and unhealthy seedlings with weak and unhealthy newly born child and his growth related problems. Similarly with high seed rate, in small space many seedlings germinate which can not have enough food for themselves, like if many children are made to sit around a plate full of food for eating what will happen? No child will get enough food, each one of them would remain hungry and grow weak. Like these the scientific facts and processes were made so intelligible. That they not only learnt, them but they started seeing reason and became logical in their thinking. In other words, they became science oriented seeing meaning to every act in the farming process. This also generated conviction in them about scientific package of practices. The farmers have resistance to the blind faith demanded by the prescriptive approach, but in the participative approach of this type they go by reasoning and understanding of scientific logics, hence no resistance in acceptance of the substitutive practices.

After analysis of existing practices, substitutive or desirable practices were explained to them again with emphasis on the scientific logic or reasoning behind them. For example, instead of 25-30 Kg. of paddy seeds, sow only 5 Kg of seeds in the same area. There will be less number of seedlings, each at a reasonable distance from the other. Your initial perception would be that there would be inadequate number of seedlings for transplanting one acre of land, but only after a week of germination, you would see that tillers had started sprouting from every seedling which may be 3 to 4 per plant after a fortnight and your perception will then change. Since every seedling has enough space to grow, they would get adequate nutrients, grow healthy and you will ultimately get 4 to 5 healthy seedlings from every grain of seed sown in the seed bed.

The farmers in this sort of interactive process used to raise their doubts and relevant questions to seek clarification. For example, if the seedlings are scarce, there would be more weed growth – and roots of the seedlings would go deeper in the soil due to which uprooting would be difficult and these would be more breakage. After solutions to these problems were discussed, they used to feel delighted. This is how the process became interactive and there was no trace of any prescription in the whole process of communication.

Under this process the entire package of practices got completely transferred with great amount of acceptability and many more farmers turning to be innovators. Hence adoption became faster.

(5) Translating communication theories into practice: In order to make the communication more effective not only to help them acquire scientific knowledge, but become more adoption prone, the communication theories

of 'source credibility', arousal of achievement motivation etc. were put to practice during the village training camps. 'How' is being explained below.

(a) Source Credibility – We know that if credibility of the communication source is high, acceptance of communication contents will also be high which will result in higher knowledge gain. We also know that credibility of the source has two components 'expertise' and 'trust'.

Now in every training camp, a subject matter specialist – a senior scientist from Agril. University or IARI used to be called to impart them training. While introducing the scientist or the trainer to the farmers, his expertise in terms of research work done, number of years spent in scientific pursuit, some special achievements made, were highlighted which gave a clear impression about his expertise. Along with this his concern and desire to see his scientific endeavour and achievement to bear fruits on farmers' fields and the satisfaction derived form this has mobilized him to come to you and share his experiences with you, was highlighted which generated trust in him. The farmers felt that he had not come merely to perform his official assignment but has come with a genuine desire to help farmers make their crop more productive. The official assignment used to be skilfully denied and converted into willing participation. These certainly highlighted the credibility of the trainer in the farmers' eyes which made the knowledge transfer process more effective. Similarly, a few words were used to convey homophillic perception to the farmers (sense of homophilli with the trainer).

(b) Motivation arousal – We know that "motive is an emotionally charged thought" which generates energy in an individual to make him move and take action or in other words 'sets the motivational process'. This used to be achieves by making a comparison with known achievers like Punjab and Haryana farmers and stressing that they are in no way inferior to them, rather they have the potential and nature's endowment (like good soil and Water etc.) to go ahead of them, they only need to accept this challenge. This is how by using simple tricks of the trade, based on our theoretical \understanding of communication process and managerial skills, the training camps were made very lively and effective.

(6) The Impact Visibility: The impact of these efforts were quite clearly visible on the farmers' fields which used to give encouraging feedback, sense of success and desire to achieve further excellence in the extension team carrying out the programme. Some of such visible impacts were-

(a) Reduction in seed rate of paddy from 25-30 kg to 2 to 5 kg (per kattha of soil bed) by an overwhelming majority of rice growers.

(b) Plantation of 1 to 2 seedlings per hill instead of 10 to 15 seedlings by the same majority of farmers.

(c) Transplantation of 3 weeks old seedlings against 50 to 60 days old seedlings as before.

(d) Balanced use of fertilizers in place of an over dose of nitrogenous fertilizers alone being used earlier.

(e) Discontinuing of home grown grains being used as seeds and tendency to procure improve seeds from outside and many farmers started scientific seed production – certified as well as truthful seeds on their own.

(f) Chemical seed treatment became a common practice.

(g) Proper plant protection measures became quite common

(h) Raised seed bed technique and strip planting methods of paddy were adopted by a large number of farmers.

(i) System of Rice Intensification (SRI) method of rice cultivation was adopted by good number of farmers.

(j) As a result of all these measures taken up by farmers paddy yield in the area increased by 150 per cent. Similar changes in other major crops like wheat and pulses were also seen.

(7) Comprehensiveness of the Field Extenuation Programme: The field extension programme was not limited merely to promotion of improved farming technology and increased productivity. It did take care of the other pressing needs for a system of progressive farming to set in. For example, availability of good quality seeds was a big problem since the state seeds corporation became defunct and the private seed dealers in the market used to sell all kinds of seeds, authenticity of which were repeatedly questioned by the farmers after having used them. This obviously generated a need of establishing a seed producing institution in the area with trust and reliability. Thus a farmers' institution called " Patliputra Beej Utpadak Swablambi Sahkari Samiti" was established in the project area and got registered under the Bihar Cooperative Society Act. The members of the samity were given training in seed production technology at institutions of repute like Regional Research Station of IARI at Karnal which has long experience of seed production, Seed Science and Technology Division of IARI and Rajendra Agricultural University Pusa, Bihar. Then certified seed production activities were started on the registered seed growers fields who were trained for the purpose. For this, foundation seeds used to be procured from ICAR institutions, Agril Universities of Andhra Pradesh and Bihar and State Farmers Cooperations. The seed farms got registered under the Bihar Seed Certification Agency and the seed production programme was initiated with strict supervision and monitoring by the project staff. The seeds produced were named "Patliputra Beez" and were given certification tag by the Bihar Seed certification Agency. Soon 'Patliputra Seed' became very popular for its purity, germination and seedling vigour, not only in the project area, but also in the neighbouring districts like Bhojpur, Kaimur, Nawada, Gaya, etc.

Another problem which the farmers faced was availability of pure (mixture free) fertilizers, and plant protection chemicals. In order to mitigate this problem, another institution called 'Kisan Seva Kendra' was established again with farmers' participation which undertook the services of input supply. This not only improved the inputs availability but also impacted the local market. Good quality chemicals and fertilizers found their places in the local dealer's stores also.

(8) Field Extension is a Continuous Process: As stated above, the crops productivity substantially increased in the project area, yet the farmers were not as well off as they should have been. Obviously then, something more needed to the done to bring the desired prosperity in the farming community. One of the idea which cropped up was that crop diversification was needed to include a cash crop in the cropping system. Perhaps this would improve their financial capability. Hence a search began, and we finally settled with introduction of aromatic plants cultivation in the cropping system since aromatic oils had good market and reasonably high price. Two aromatic plants were selected for cultivation.

(a) Japanese mint – This fits very well in rotation with rice crop. Since majority of rice fields are low lands, high soil moisture persists till mid December, making wheat showing much delayed which adversely affected wheat yield. But sowing time for mint is January (direct sowing) and even February for transplanted crop. Hence this fits bitter than wheat and even with one cutting by March – April, the oil yields of mint generate much more money than the sale proceeds of wheat crop. After April the land is again available for paddy plantation. With experience, a more intensive cropping system was discovered – Rice - Wheat – Japanese Mint. After rice early maturing wheat is planted and in January seedbeds of mint are sown to produce more than 1 month old seedlings by March when wheat is harvested and then the field is given flood irrigation and mint seedlings are transplanted without giving any cultivation to the wheat field. The mint crop gives one good cutting by April or early May and leaves the field for paddy cultivation. This substantially increased cash flow in the farming system and provided desired liquidity to the farmers.

(b) Lemon Grass: There are some upland in the area which are not very productive. Generally only one crop in a year is taken. Such lands rae ideally suited for cultivation of lemon grass which is a five year crop and gives regular income at three months intervals. These two aromatic plants cultivation was introduced in the area and a distillation plant was installed at the project head quarters for extraction of oils. This caught up very well and now five more distillation plants have been installed at various locations and the cultivation is spreading quite fast. This is going to create a lot of economic activities and employment generation in the near future. With this project in action for which financial and technical support have been provided by TIFAC, a council under Department of Science and Technology, Govt. of India, the Paliganj area of Bihar is on move. Transformation not only in

Some project impacts at a glance:

1) Saving in paddy seeds

   Local practice -25 kg seed for 1acre planting

   Replaced practice – 5kg seed for 1 acre planting

   Saving of seed – 20kg seed per acre or 50kg seeds per hectare

   The project area has 11,000 hectares under paddy crop. Hence, total seed saving is 11,000× 50kg =5, 50,000kg or 550tons.

   In terms of cost –

   1 ton seed costs Rs. 15,000/-

   Hence, 550 tons of paddy seed would cost Rs.82, 50,000/-

   Hence, Rs. 82, 50,000/- worth of paddy seeds are saved

2) Productivity of paddy on an average has increased from 20.13 quintal per acre to over 40 quintals per acre which is reflected in the F.C.I. procurement of paddy seed made at Paliganj in 2007-08. This year F.C.I. has procured 72,000 Quintals of paddy from farmers of the project area as against 15,000 quintal procured in 2005-06.

3) 31.8 tons of certified paddy seeds were produced in 2008, which is worth 31.8×1500 = Rs 58, 5000/-

4) Due to higher productivity and farm income credit worthiness of farmers of the project area was recognized by the local commercial banks which resulted in providing kisan credit cards to 2379 farmers for total credit limit of Rs. 7,75, 000.00/-

Impact Outside the Organization

(i) There is a huge demand of the certified seeds produced under the Pataliputra Beez Utpadak Swabalambi Sahakari Samati from farmers of neighbouring districts of Bhajpur, Kaimur, Nawada etc. and in 2007-08 the project supplied to them 7,633kg of certified paddy seeds.

(ii) Farmers from neighbouring districts like Gaya, Jahanabad, Aurangabad, Arwal and Begusarai continuously approach the project to impart them the knowhow for cultivation of aromatic plants and demand planting material. The project has responded to some such demands for man and material resource.

(iii) The technology adopted by the farmers in the project area has crossed the project boundary through farmers to farmers' contacts.

(iv) A two-day Krishi Prasar Mela is organized every year in the project area. Farmers from outside the project areas visit the mela in large number, take Part in the scientists –farmers discussion held in the mela, make purchase of farm input like seeds chemicals, farm implements etc. and update their knowledge with various displays made in the mela.

agricultural system but in the rural economy as a whole has started taking place. His Excellency Dr. A.P.J. Abdul Kalam ex- President of India visited this project three times, and Shri Nitish Kumar, Chief Minister of Bihar visited it once and both of them desired that this effort must be multiplies in other parts of the state and the nation to bring prosperity to the rural areas.

Chapter 17

Extension Education Perspectives in Changing Agricultural Scenario

C. Prasad

Background

After Training and Visit extension Project of the World Bank, since 1985 the extension education system in agriculture has come to a cross-road – the agricultural extension system of the Ministry of Agriculture, the extension system with major responsibility of Transfer of Technology (TOT) in the country. The ICAR/SAU extension system plays the first-line extension role: **(i)** bringing latest farm technologies promptly to the notice of **(a)** a limited progressive farmers of the states/districts; and **(b)** all the state extension functionaries. While the ICAR/SAU extension system has continued mainly through its KVK project, but the state extension system was weakened due to resources of T and V system being withdrawn – in fact, a vacuum has been created which needs immediate attention and reforms.

As a matter of fact, our extension system since 1952 – the era of planned development, continued with the foreign supports – C.D., AES, IADP, NYVP and T and V, so much so that even the name of national extension system could not be stabilized. Lately, it was proposed to the government at the highest level as to why we do not have our own extension system – the name does not cost anything, but gives stability and sustainability of the system. If not anything else, it may be conveniently called, the "National Agricultural Extension System (NAES)." Perhaps, this has been accepted by the Ministry of Agriculture, Govt. of India in a subtile manner. The USA extension system was named as the "Cooperative Extension Work" since 1914 and still it is the same. We adopted foreign projects with their names and resources and forgot about our own.

The vacuum created by the T and V system, could not, perhaps, be perceived by the government which functions on its administrative strength, not on technical and intricate educational system. Extension is an informal and out of school education – a more difficult concept; Ministries, by and larges, continued with the agricultural extension more offering agricultural extension services than education.

Now the ATMA (Agricultural Technology Management Agency) at the district level and SAMETI (State Agricultural Management and Extension Training Institute) at the state level are being prompted only in half the districts in the country (252 out of 600).

The functional linkages between the ICAR/SAU system and the state extension mechanism remained always a new subject, for linkages and convergence occur when more than one system have **(i)** common goals and interests; and **(ii)** they are equally committed to the goals. Do such conditions exist anywhere in the country ?

Conceptual Framework – Policy Issues

Defining Extension

The basic discipline for TOT is Extension Education which when applied to agriculture is Agricultural Extension, to Veterinary is Veterinary Extension or to Industry is Industrial Extension. Education is the keyword per cent all TOT efforts, no matter who does it (ICAR/SAUs/State Departments, etc.), are educational in nature/character: education, training, demonstrations, fairs, exhibitions, use of mass-media, information technology – all are ment for education and motivation of the farmers which may lead to acceptance of recommended technologies, and then finally the adoption by the farmers as well as diffusion in the farming communities.

Agricultural extension services are inputs support (seeds, fertilizers, chemicals, pricing policy, crop insurance, credit policy, land reforms, state farms management, soil and water testing, etc.), are the normal responsibilities of the State Governments, in addition to the extension education function. In reality, the state functionaries are doing mostly the extension services rather than playing the educational and motivational roles.

Nomenclature of Extension System

Since planned development in 1952, the agricultural extension system (s) has been named, by and large, after the sponsored programmes of the foreign development agencies or their influences – C.D., AES, IADP, IAAP, HYVP and T and V. We do not have our own national nomenclature even after 60 years of independence. Naming a system does not cost anything, but gives stability and sustainability to a system. In the USA, the extension system is functioning under the banner of "Cooperative Extension Work" since 1914; all programmes, strategies, projects or extension sub-systems work under this umbrella, and changes are made as and when needed under this banner.

The question is What is our extension system ? Does the note of the Ministry of Agriculture makes any point on this issue ? No, so why can not we name it as the National Agricultural Extension System (NAES) or, for that matter, any other suitable name? ATMA (Agricultural Transfer of Technology Management Agency) was originally launched as the district level management agency for integrated working of varied extension sub-systems. ATMA, therefore, is an national level approach, programme or strategy; not a national level extension Organization. An organization would encompass dimension every of development, may that be policy, strategy, investment, governance or WTO?

Resources Distribution

As per the policy in the ICAR/SAU System, 10 per cent of the total budget and 20 per cent time of the scientists are expected to be spent on TOT. This is a reasonable proposition, but are they happening? If we add to this, the extension education role of the Ministry of Agriculture/State Governments (exclusive of services role), the budget, normally should be more than one-forth of the overall total agricultural budget?

Extension Functionaries

There are two categories of staff who are directly involved in Extension Education Programmes like the two sides of the coin:(i) Extension Education Professionals and workers; and (ii) the Subject-Matter Specialists (SMSs) – Agronomists, Horticulturists, Entomologists, Animal Scientists etc. This dimension of TOT has to be understood by both the professionals per cent one can not do without the other. At present, the common understanding is that TOT means the job and responsibilities of the extension professionals only or if at all, the SMSs are the light support to extension programmes without owning it as their own programmes or responsibilities. This dichotomy has to be done away with so that both feel responsible to the extension programmes and they together share the credits or discredits accruing out of these activities. Staffing of the TOT System, therefore, has to be planned accordingly. Both in fact should form a Cadre of TOT in the country as a long-term strategy; then their future is known and their commitment to the system could be ensured; their expertise and experiences could be utilized within the Cadre.

Professionalism in Extension

Our professional goal is to bring about 'professionalism' in the discipline, which is defined as 'reaching the altar of functionalism from where once can see the goodness of fit as well as the maladies, if any. Profession has a theoretical foundation leading to practical acts and contributions. Dictionaries define professionalism as the quality

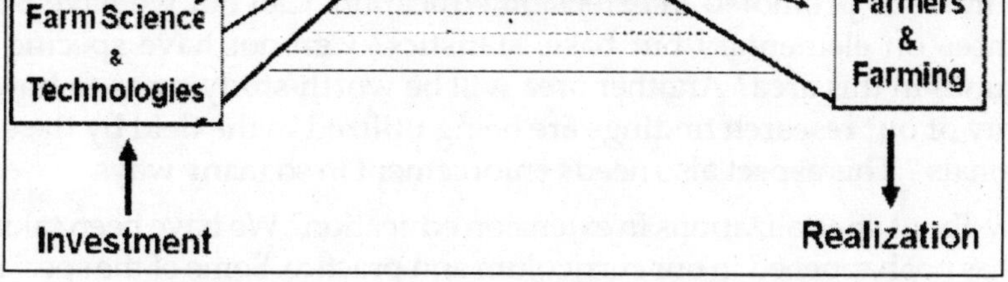

Figure 17.1: Extension Education – The Bridge.

of a profession or discipline. Professor is one who has strength and ability to profess – one who can declare freely about the right and wrong of his discipline based on his vast studies, rich experiences, strong-convictions and futuristic ideas and projections like a astrologer. Professionalism is a religion based on the acts of 'work is worship.'

Human Resource Development (HRD)

Once I was reading the Reader's Digest in a train when I found two articles very educative: first talked about Truth and truthfulness – after his class was over, the teacher wrote on the black board: Satyam Bada. In the next morning, he came to the same class-room, he found written below Satyam Bada, but Mr. B. Lal is more successful ! What example Mr. B. Lal presented before the students? Then the second article was by our former Prime Minister Shri V.P. Singh; he wrote that elders have responsibility to our younger generation by putting good examples?

What are our curriculum in extension education ? What kind of teachers and trainers we have? How could we produce excellent extension scientists and workers? This is a Herculean task and our extension education societies/congress should pay attention to these basic issues besides educational institutions.

The ratio and relationship of extension education contents in the curriculum and the subject-matter contents by way of minor(s) should be examined. Our extension with their social engineering proneness must have at least two broad areas: (i) at the knowledge and skills in their own specialization; and (ii) the knowledge related to the recommended package of technologies. Only on the occasion of specialized subject-matter demands, that the subject-matter specialist should be invited; that much proficiency we must attend. And this could be possible only when on basic minor subject-matter courses, we keep on organizing refresher courses as a design for the field extension scientists as well as the field workers.

In addition, training courses, as a policy, could be designed for the subject-matter specialists on extension approaches, methodologies, strategies and management. The SMSs should be given more opportunities for Seminars and Conferences on extension education, whereas the extension professionals should be given more opportunities in the subject-matter Seminars and Conferences.

Our research programmes including post-graduate researches must be occasionally reviewed, collected, collated and documented. As has been reported and observed, the standard of research in extension education has been repetitive and has been weakened. This needs examination. Besides, it has also to be realized that for objectivity in our research, our background of basic statistic relevant to social sciences must be promoted in extension education. Can not we have Seminars/ Conferences on elementary but basic statistics? Can not have specific training programmes in this area? Another area will be worth-studying as to how far and how many of our research findings are being utilized in the field by the extension professionals ? This aspect also needs enforcement in so many ways.

How about specializations in extension education? We have been taking about it, but it has not happened in our curriculum and practice. Some of the specializations could be: (i) approaches, methodologies and strategies; (ii) extension education organization and systems and management; (iii) information and communication technologies; (iv) training and entrepreneurial development; and (v) research methodology.

There has been a lot of gap between Saying or Writing and actually doing in the fields; our management proneness and commitment have been questioned. We must be alive to these vernability. Our societies and extension education Forums/Congress/State chapters must devote some time to these critical issues.

Skewed Development

After over 60 years of Independence, we are still struggling on poverty and hunger for at least over 25 per cent of the rural population; they are deprived, vulnerable and handicapped. The salvation has to come from agriculture and cottage industries including agro-industries. So far there has been skewed development in the country focusing on economic gains, consequently the social side of development has not only been ignored, but almost forgotten. This has led to "mental poverty" among well-to-do people – greed for wealth and power leading to corruptions and chaos in the society. Leadership, which should drive people to prosperity, peace and salvation, is by and large self-centered. Such trend has to be arrested by bringing the positive social values in leadership and people in general, and development professionals and social workers in particular. A balancing act has to be done between economic and social development very seriously.

Nobel laureate Amartya Sen has so definitively demonstrated that empowerment leads to entitlements and entitlements lead to enrichment. The skewed distribution of the fruits of the impressive progress the country has made, requires immediate course correction because economic progress in itself cannot be our national goal.

Holistic Approach and Methodology

- ☆ Planning in the country has been more top-down than bottom-up – a demand-driven planning, whereas planning literature has always talked of planning from below involving all target groups, interest groups and stakeholders. In reality planning in India has been done by more scholars and academicians than field workers and realists; planning by those target groups will help them believe that the plans are theirs, which will help in their acceptance, implementation, adoption and diffusion in the communities. In holistic approach down to earth planning is crucial.
- ☆ AMUL Model – In the past, we have given more thrust to crops than other sub-sectors – horticulture, animal husbandry, dairy or fisheries. Then our extension system has been half-backed – went up to raising productivity and production ignoring storage, grading, processing and marketing – the total process A to Z, bringing more income and employment to farmers. The living example is the AMUL/NDDB system of holistic extension: organizing dairy farmers, arranging inputs – money, animals or allied material support, pricing, collection/purchasing, processing and marketing. This is the best example in the world of a holistic strategy.

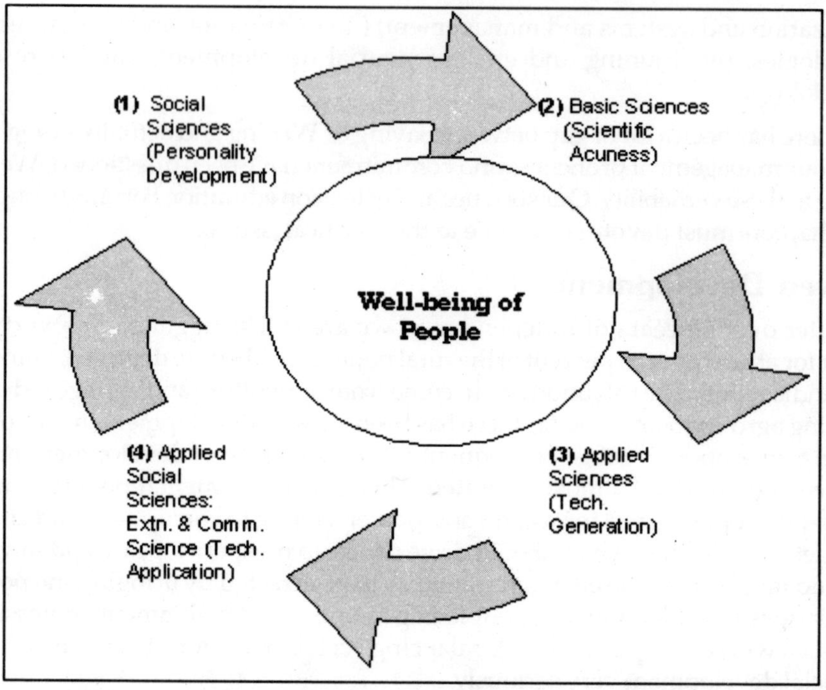

Figure 17.2: Theorizing Social Sciences – Social Sciences First.

- The FELDA Model of Malaysia: The FELDA model of eradicating poverty of Malaysia is an extremely useful model for the poorest of the poor especially the landless and the homeless. The Malaysian Government, for instance, identifies big piece of land, develop it fully including levelling of lands, creating irrigation facilities–tube wells, irrigation channels, tractors etc.; and also constructs modest houses for the poor at the same site as well as develop a marketing Centre for them. The selected poor farmers then are given the lands and houses at the site on the long-term soft loans after a certain initial years of grace. We were shown that such farmers / rural people within a decade had another set of houses constructed on their own, had motor cycles, good education to the children, very happy community living, etc. For landless and houseless, there cannot be a better development model really. In India, we have always taken a truncated view of development for the poor – they were given a poor piece of land under Vinoba movement without any initial support to develop it or inadequate money for the housing. It has been all along a half-hearted approach – we are hardly seriously interested in the poor families / people it seems. Then there are leakages of resources ment for the poor – Late Hon'ble P.M. – Shri Rajiv Gandhi said, "only 18 per cent of the resources ment for the poor reaches them"
- The agro-industrial complex of Bulgaria: In 1976, there were 119 such complexes in a small country like Bulgaria. This scheme is based on

horticultural products especially vegetables and fruits. Under this model, those who have been growing vegetables and fruits have been organized, in the first instance into clusters and complexes/forums; their assets and resources have been studied and known; they have been given a remunerative prices of the vegetables and fruits right in the beginning for the year as a whole; the required inputs in terms of credits or farm inputs have been assured; the products have been collectively processed; and then marketed both in domestic markets as well as in export markets. The farmers of the complexes benefited by remunerative prices all the year round.

☆ SGSY (Swarnjayanti Gram Swarozgar Yojana): The latest design of rural development has come from Bangladesh – the Self-Help Groups (SHGs) formation and credit and thrift schemes – also known as micro-enterprises and financing. It has started with empowering the rural poor women: organize them in a group of 10-15 members, training them about SHGs; orient them to have a regular monthly meeting – get-together; develop a habit of small savings (Rs. 10.0-15 or 20) per month; select the Chairwomen, Secretary and a treasurer; treasurer keeps the money so collected; then start giving loan to the members on an interest collectively decided by the SHG; and when they have more money, say Rs. 5,000.0 or more, then they can open a bank account; and after a while, when they have 10-15 thousand collected in the bank, as per the policy of the government, they can take loan four times the deposited money for productive purposes – any small enterprise, shops, printing press, Agarbati-making, candle-making, goatary, piggery etc.

The SGSY is again a holistic scheme of the Ministry of Rural Development which covers the promotion of SHG, providing them vocational training, giving them loans if needed, some incentives in terms of subsides etc. if well done. At the cost of the government, the poorer sections of the rural people can take full advantage. Earlier it started with women empowerment only, but now it is open for all men or women. This has also emerged as a holistic design. We must use it for the farmers and farming; all extension programmes should be based on this foundation.

Agricultural Management

The National Commission on Agriculture (NCA-1976) recommended for a National Institute/College for Administration and Management as well as state level Management Institutes. The National Academy for Agricultural Research Management (NAARM), Hyderabad came at the national level, but most of the state level institutions did not come up. To quote the Planning Commission, GOI: 'The issue of governance has in the recent times emerged at the forefront of the development agenda. Good governance is one of the most crucial factors required if the targets of the Tenth Plan are to be achieved. Experience shows that while good governance can help secure human well-being and sustained development, it is equally important to recognize that poor governance could well erode the individual capabilities as well as institutional and community capacities to meet the needs of sustenance' (Planning Commission, 2002-07).

Reforms Needed in Extension System

☆ Extension system can remain the same, but the strategies must change to suit the typical environment – NEH states are different where uniform strategies can not work out. So far we did not have even the name of our extension system; we went by the foreign supports and their nomenclatures – C.D., NES, IADP, T and V, etc. A definite name does not cost anything, but gives stability and sustainability. A suggestion was made to higher authorizes to call it a "National Agricultural Extension System (NAES)", which seems to have been accepted, but consistency and continuity has to be assured. This requires concerns and commitment which seems to be in short supply.

☆ The basic and drastic change in the system demands a 'Council for Transfer of Technology in Agriculture' (COTTA) at the Centre – an autonomous body for aggressive extension leadership under the Ministry of Agriculture. This could be easily done by integrating the Directorate of Extension of the Ministry of Agriculture, Government of India, and the Division of Extension of the ICAR. These two subordinate Offices do not have their 'teeth and voice'. The Council would have virtually no financial implication when the resources and staff of both the system/sub-systems are brought together, integrated and rationalized (Annexure-I). The two arms of this Council – State Extension Education and the ICAR/SAU Extension System may continue with some modifications: the 'extension education' component (non-existent at present) and the 'extension service' component of the States must be separated; they can supplement each other but as independent functional units with their own objectives and methodologies. Extension education professionals must not indulge be in regulatory – routine functions. Organizational arrangement may be seen at Figure 17.3.

☆ We adopted the Land Grant College/University model of the USA where teaching, research and extension are fully integrated; government gives the resources and management support. This system has functioned successfully in the U.S.A. since 1914 – the Cooperative Extension Work. However, in India we accepted the truncated model of the U.S. System: (i) the ICAR/SAU Extension System was to play the limited first-line extension role to be followed by (ii) the main extension role of the Ministry of Agriculture and the State Departments of Agriculture. This approach was conceived when there was to be only one Agricultural University in each State so that their research and academic programmes might not suffer due to the heavy load of extension work and responsibilities. But the situation has changed drastically – there are more than one Agricultural University in most of the States irrespective of the Model Act of the ICAR – the Maharastra State has five SAUs? Under this condition, the area jurisdiction being reduced, the fully integrated model of the USA comprising teaching, research and extension is fairly possible.

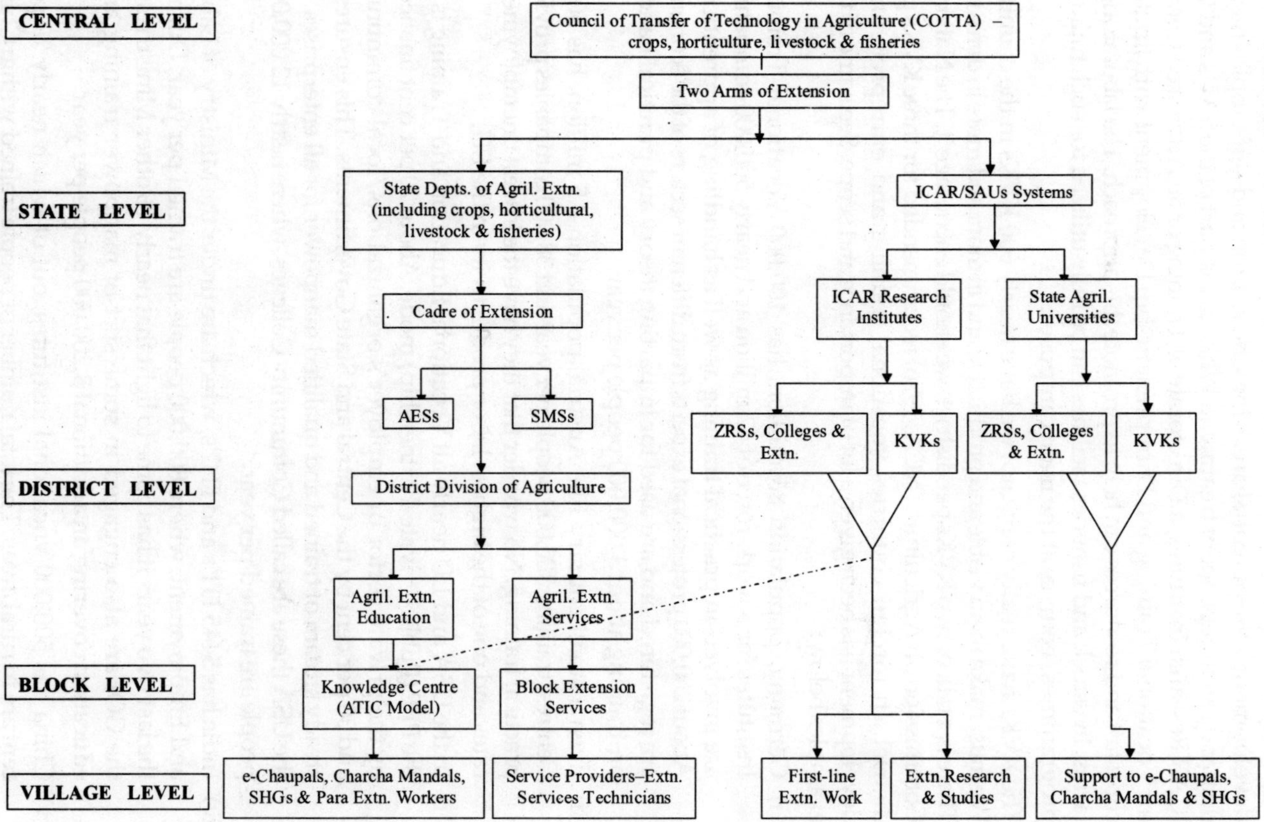

Figure 17.3: Broad Organizational Extension Structure Proposed for the Country.

☆ 'Training' is another important component of development – so far an integrated part of the extension system. But in view of its scope and importance, especially for vocational areas and entrepreneurship development – the potential areas for job-creation and self-employment, demands an independent treatment. While being part of the SAUs and the ICAR Research Institutes, a Directorate of Training or an Advanced Centre for Vocational Training and Entrepreneurship Development with limited core staff and resources can help to promote this approach. The laboratories, farms, livestock and fisheries facilities can be fully utilized for such training programmes/courses at the main campuses.

☆ The KVKs have really come up; we have nearly 600 KVKs in the country. We must make them viable and productive and in more and more in demand. There is a talk of two KVKs per district; we should encourage it. The National Commission on Agriculture (NCA, 1976) recommended for three KVKs per district. In foreign countries vocational training and entrepreneurial development has been given a lot of importance and scope. See some of the examples below:

i) Germany, population 86 million, has 100,000 Vocational Training Institutes for a work-force of 42 million and nearly 500,000 companies are involved in practical training as well as handling of apprentices. About 200,000 professional experts from different sectors of the economy are registered and are used for imparting theory and practical hands-on-training about 3,000,000 people per year.

ii) Even a small country like Austria, population 8.5 million, has 5000 Centres, training 300,000 people per year and 38,000 companies provide practical training. No wonder that they have the lowest unemployment rates and one of the highest labor productivity in the EU.

iii) In the USA and EU nearly all Vocational Education and Training is in the PPP (Public-Private-Partnership) mode. About 70 per cent financed by the private sector or employer's organizations/local community and 30 per cent by the Central and State Governments. This ensures a steady stream of trained and qualified manpower for all enterprises. In the USA these are called Community Colleges, where nearly 12,000,000 people are trained per year.

iv) India has 5445 ITI's and ITC's, which are under the Ministry of Labor and Employment, where 800,000 people are trained per year. During the last two years it has come to light that nearly 16 other Ministries of the GOI are also engaged in some sort of manpower training and education covering an additional 3,200,000 people per year.

v) China has 500,000 vocational institutes, out of which nearly 70 per cent are in rural areas. The total number of people trained within China is about 60,000,000 people per year. It is therefore not surprising to note that China with arable land much less than India produce's nearly 100 per cent more food grains, although it has a very harsh climate.

Chapter 18

Farmers'- Led Extension: Experiences and Road Ahead

K.D. Kokate, P.B. Kharde, S.S. Patil and B.A. Deshmukh

The present day agriculture is defined by key concept of stability, sustainability, diversification and commercialization. In the last decade, the agricultural situation in India has undergone a tremendous change in the light of liberalization and establishment of World Trade Organization (WTO). India's signing of General Agreement on Trade and Tariff (GATT) in 1994 and joining of WTO has put our agricultural into a framework of global market. Low productivity of crops added to less remunerative market prices of agricultural commodities are the major causes of worry. Thus, agricultural enterprise is found to be not very profitable although a large majority is depending on it. With the globalization of agriculture, major emphasis is laid on increasing the productivity of crops. During the last 50 years, major emphasis has been given on Production-led Extension (Duraisamy, 2007).

In the changing situation, farmers need to transform themselves from mere producer-seller in the domestic market to producer cum seller in a wider market sense to realize good returns on their investments, risks and efforts. In order to achieve this goal, farmers need to know answers to questions like what to produce, when to produce, how much to produce, when and where to sell, at what price and in what form to sell their produce. Farmers have received most of the production technologies from the extension system. The extension system needs to be oriented with knowledge and skills related to the market. This revamping of extension system will certainly play a catalytic role for ushering in farmer-led and market-led extension in India (Moni, 2004).

Need of Paradigm Shift in Agricultural Extension System

India's agricultural extension system is at a pivotal point in its development. The arrangements for agricultural extension in India have grown over the last five decades, in terms of activities, organizational types and available manpower. At the outset, extension worked to bring about broad-based rural development. However, the food crisis starting in the late 1950s refocused the efforts of extension on food security and increasing food production. The combination of Green Revolution technology in the late 1960s and Training and Visit (T and V) system in the mid-1970s enabled India to achieve food self-sufficiency during the 1980-1990s. Thereafter, the Government of India with assistance of World Bank designed and pilot-tested a new extension approach *i.e.* Agriculture Technology Management Agency (ATMA) that decentralized extension system, focused on agriculture diversification, thereby, that making it more market-oriented. This market-driven extension system *i.e.* ATMA has really made an impact in increasing farm income by organizing the farmers and making the system viable through bottom up planning (Singh *et al.*, 2006).

While discussing the various issues for agricultural extension of 21 th century in India, Samanta (1991) highlighted the importance of institutional reconstructing and renewal, and decentralized extension structure. Recently many developing countries have reaffirmed the essential role that agricultural extension can play in agricultural development as pointed out by Birner *et al.* (2006) and Anderson (2007). Therefore, extension functionaries need to play a major role to build the capacity of the farmers to meet the emerging challenges and make the farmers to realize better prices to their farm produce. Based on resynthesis of framework developed by Zijp (1995) and River (1997),who observed that extension has to once again reorganize as a service for providing information, advice and education, rather than technology transfer. This renewed definition of extension multiplies its potential to move beyond purely production and productivity concerns to those knowledge and information intensive, organizing clients into groups and marketing. In order to successfully address above issues and challenges, extension need to be looked into an altogether new perspective. The farmer-led and market-led extension so far is a peripheral issue and alternative approach in today's extension scenario. Hence, there is an urgent need to shift the paradigm of extension system from mere production-led to farmer-led and market-led extension system on end-to-end basis. Therefore, new framework for conceptualizing and operating extension system is evolved and diagrammatically presented in Table 18.1 and its practical approach has been discussed with relevant experiences.

MPKV Experiences of Farmers'-Led Extension Approach

The Mahatma Phule Krishi Vidyapeepth (MPKV), Rahuri has been implementing the novel concept of Farmers-Scientists Forum (FSF) since October, 2005. The Vice-Chancellor is president, while, the Director of Extension Education is Executive President of this FSF. The FSF established at Central Campus, comprises 40 Farmer-Scientist Clubs (FSCs) of 1500 member farmers in the jurisdiction of ten districts of the university. Each FSC has coordinator and these FSCs are attached to the Colleges, Agril. Research Stations and Agricultural Technology Schools. These FSCs function

Table 18.1: Paradigm Shift from Production-Led to Farmers-Led Extension System

Components	*Production-Led*	*Farmers-Led*
Purpose/Objective	Transfer of production technologies	Capacity building (especially farmers extensionist), create para-professional extension workers,creating or strengthening local institutions
Goal	Food self-sufficiency	Livelihood security including food, nutrition, employment to alleviate poverty, sustainability and conserving bio-diversity
Approach	Top-down, commodity and supply driven	Participatory, bottom-up and demand driven
Actors	Mostly public institutions	Pluralistic with public, private, non-government and farmers organizations as a partner rather than competitors
Mode	Mostly interpersonal/individual approach	Integration of clients oriented on-farm participatory/experiential learning methods supported by ICTs and media
Role of extension agents	Limited to delivery mode and feedback to research system	Facilitation of learning, building overall capacity of farmers and encouraging farmers experimentation
Linkages/liaison	Research- Extension-Farmers	Research- Extension-Farmers Organizations(FIGs, CIGs, SHGs)
Emphasis	Information management, Production "Seed to Seed"	Knowledge management and sharing
Nature of technology	Input intensive, crop based and general recommendations as per agro-climatic zone, fixed package of information	Knowledge intensive, broad based, farming system perspective and blending with ITKs.
Critical areas	Improvement, production and protection	Decision support system, integrated farming system approach, natural resource management, clients group formation and community empowerment
Critical inputs	Money and material	Access to Information, building human and social capital
Accountability	Mostly government	To farmers rather than donors

as Commodity Interest Groups (CIGs). The basic idea of formulating the FSCs is to demonstrate and to educate the farmers about the technologies generated by university at different locations. MPKV has developed the farms of these members as a model for the village and these farmers in turn are servicing as farm leaders to disseminate the university technologies to other farmers in their area. Participatory research- extension approach is being promoted through the FSF. Regular Monthly Meetings and Review Meetings are organized, wherein there is interaction between FSC members and university scientists. Database of soil, water, cropping pattern, various enterprises of member farmers etc.has been developed by the university and accordingly location specific interventions are suggested for increasing their income through agro-based enterprises. Some of the experiences of members of FSF are depicted below:

- ✰ Seed Village Programme are regularly implemented on organized farms of members of FSCs. MPKV implemented such programme in Varur village, Dist. Ahmednagar on gram (Variety: Virat) on 70 acres in the year 2006-07. The village produced 430 quintals of Virat seed worth Rs. 21.50 lakhs, which really boosted the socio-economic condition of the farmers.
- ✰ Another seed programme in Tambhere village of Ahmednagar district in 2007-08 yielded 12 farmers an amount of Rs 2.48 lakhs through sale of 82.8 quintals certified seed of gram variety Digvijay.
- ✰ Demonstrations on yield targeting approach (250 tonnes/ha) of Co-265 variety of sugarcane released by MPKV was carried out on the farms of members of FSCs in Pune and Ahmednagar district in 2007-08. These farmers realized average yield of 215 tonnes/ha.This sugarcane variety is now gaining prominence in the sugarcane tract of Maharashtra through successful FLDs.
- ✰ Capacity building of 800 members of FSCs has been carried in 2008-09 through the programme on Scaling up of water productivity of increasing livelihood through teaching cum demonstration to the farmers and trainers, a project sponsored by Ministry of Agriculture, Govt. of India, New Delhi.
- ✰ Demonstrations through Farmers Participatory Action Research Project sponsored by Ministry of Water Resources, New Delhi are being implemented on micro-irrigation systems on farms of 100 member farmers of FSCs in the university jurisdiction since 2008-09.
- ✰ Good example of farmers'-led and market-led extension has been created through formation of FSCs. A woman member farmer of FSC Mrs. Vanita Gunjal from Kandali village of Pune district produces continental vegetables for marketing in five-star hotels in metropolitan cities like Pune and Mumbai. She owns meagre land of 1.87 ha., but the satellite marketing under the guidance of university scientists have reaped her rich benefits. Marketing of these exotic vegetables is done under the trade name of *Green Pallavi Farms and Sales.* Mrs. Gunjal is the winner of N.G.Ranga Farmer Award for Diversified Agriculture (2005-06) of the ICAR, New Delhi.
- ✰ Another member farmer of FSC Shri. Hanumant Gajare from Aran village of Solapur district has put forth an ideal example of farmers'-led/market-

led agriculture through bumper pomegranate production in scarcity area. He has cultivated Phule Bhagawa pomegranate for export purpose on 1.20 ha. area. This farmer realized a total income of Rs. 37.20 lakh through pomegranate export to European countries in 2008-09 season, whereas, he incurred an expenditure of Rs 3.0 lakh. Apart from these examples, the innovative approach of Farmer-Scientist Forum has led to Farmer to Farmer extension in FSF has helped in reduction in cost of cultivation and increased monetary returns through value addition and marketing. These member farmers of FSCs have achieved outstanding awards at national and state level. All these findings are in line with those of Wakle *et al.* (1998), Bellurkar *et al.* (2003) and Ekale *et al.* (2003).

Strategies for Promoting Farmers'-Led Extension

- ☆ Promoting Farmers Interest Groups, Women Interest Groups, Commodity Interest Groups for strengthening Farmers'-led extension.
- ☆ Capacity building of these organized groups for advanced agricultural production technologies.
- ☆ MPKV model of Farmers-Scientists Forum need to be replicated on a wider base.These member farmers will serve as para extension workers in their social system.
- ☆ Promotion of farmers participatory research and extension programmes for location specific technology development,refinement and dissemination.
- ☆ Farmer groups to be equipped with market intelligence and Information and Communication Technology (ICT) for access to market information.

References

Anderson, J.R., and G. Feder. 2007. Agricultural Extension. In R.E. Evenson and P. Pingali, eds. *Agricultural development: Farmers, farm production and farm markets*, vol. 3 of *Handbook of Agricultural Economics*, Amsterdam: Elsevier 2343–2378.

Bellurkar,C.M., P.K.Wakale and M.A.Gholve.2003.A study on decision making pattern and participation of rural women in animal husbandry and dairying enterprise. Mah.J.Extn.Edn. 22(2) : 81-85.

Birner, R., K. Davis, J. Pender, E. Nkonya, P. Anandajayasekeram, J. Ekboir, A. Mbabu, D. Spielman, D. Horna and S. Benin. 2006. *From best practice to best fit: A framework for analyzing agricultural advisory services worldwide*. Development Strategy and Governance Division, Discussion Paper No. 39. International Food Policy Research Institute (IFPRI), Washington, DC.

Duraisumy, D.G.2007 : Market Led Extension : Emerging Perspectives. In F.M.H. Kaleel, Jayagree Krishnonkutty and K.Satheesh Babu (eds). Market led Extension -Dimensions and Tools. Agrotech publishing Academy, Udaipur : 42-51.

Ekale, J., D.M.Nikhade and C.M. Bellurkar, 2003. Role perception of farm women in farm activities. Mah.J.Extn.Edn. 22(2):90-93.

Moni, M. 2004: Ushering Market-led Agriculture Extension. Available at http://www.i4donline.net

Rivera, William M., 1997. Agricultural Extension in the next decade. European Journal of Agricultural Education and Extension, 4(1): 29-38

Samanta, R.K.1991. Agricultural Extension in changing world perspective. Uppal publishing House, New Delhi : 137-141.

Singh, J.P. Swanson B.E. and Singh K.M. 2006. Developing Decentralized, Market-driven Extension System in India : The ATMA Model. In : Van Den Ban A.W. and Samanta R.K. (eds) Changing Role of Agricultural Extension in Asian Nations, B.R. publishing corporations, New Delhi, pp : 203-223.

Wakle, P.K., V.T. Wattamwar and M.I. Khalge 1998. Utilization of different sources by farmers for seeking farm information. Mah. J.Extn. Edn.17:299-300.

Zijp, W. 1995: Personal communication. Unpublished paper.The World Bank, Washington, DC.

Chapter 19

Extension Perspective in Changing Agri-rural Environment

P. Das

The Millennium Development Goals Adopted by the International Community - 2000

First Goal to eradicate extreme poverty and hunger

Eighth Goal: to build a global partnership for development to make available the benefits of new technologies, especially in information and communication

In the context of growing public concern for the environment, globalization, household food security and eco-regional imbalances. We need to have new directions in planning transfer of technology by designing more effective linkages between scientist and farmers for technology information to meet the future challenges of growing demand of food production with egalitarian consideration.

Two Models of Diffusion of Innovations

A. In Homogenous Population (With equal resources)

B. In Heterogeneous Population with Differential Access to Resources.

Agricultural Challenges – Fourfold

Need for enhanced production and productivity to feed the growing population and to retain the economic viability of agriculture as a profession

Need to address issues of equality and uneven development: so that gains of growth are more equally shared across regions and peoples

Table 19.1: Agriculture Extension in Post-Independence Era

1950s1960s	Farmers' Ignorance	Extension Education	Teaching	Diffusion adopters laggards etc.	Questionnaire Surveys	Diffusion Research
1970s1980s	Farm-level Constraints	Removal of constraints	Input supply	Constraints farming system	Questionnaire surveys, on-farm research	Farming system research
1990s	In-appropriate technology	Farmer participation	Facilitation	Participatory approaches and methods	Discussion, observation, diagramming by and with farmers	Farmers participatory research, Farmers-First, PRA etc.

Need to understand and address issues of sustainability: environmental, production, institutional and fiscal.

Need to enhance profitability in agriculture: in the light of liberalization process and open up international markets.

Knowledge Drives Development

- ☆ We used to think of capital as the scarce factor in production
- ☆ Transfer of capital as the key instrument of growth
- ☆ Knowledge is now a important factor in development
- ☆ This trend is set to intensify
- ☆ Knowledge accumulation and application will drive development processes and will create unprecedented opportunities for growth and poverty reduction

The Web is The Largest and Richest "Agricultural Information System" Discovery Of W3, WWW, World Wide Web In CERN (*High Energy Physics Lab*) Lab, Geneva, Switzerland. Internet protocols like Hyper Text Transfer Protocol (HTTP) and Hypertext Mark-up Language (HTML).

- ☆ Cyber farm for the use of local farmers of Illinois
- ☆ Stratsoy, the first major commodity portal
- ☆ Companion website of Illinois Agronomy Handbook including climatic information over 100 years
- ☆ FARM.DOC viewing expert information on grain markets and other commodities including Expert FAQs, tool suites and pricing information funded by Illinois
- ☆ Soybean Board
- ☆ FARM.edu project, an integrated cross-disciplinary farm management system for small to medium-sized farmers
- ☆ FARMER Business: Systems
    1. Research institutions
    2. Public administration
    3. Farmer
    4. Open market/global scale
    5. Unlimited partners (known and unknown) Static content

Interactive Voice Response System: on Local Weather Forecast, and Market Information

Sources of availability of inputs

Online consultancy to farmers

Other services related to WTO:

- ☆ Product Information Guide
- ☆ Fundamentals and vocabulary
- ☆ Quality management requirement
- ☆ Quality reporting procedures
- ☆ Certification procedures

Dynamic Content

Data base on Frequently Asked Questions (FAQs)

Medium range weather forecasting and crop/animal care

Food Standard

- ☆ Argentina proposes a regional Latin American food code, *Codigo Latino Americano de Alimentos.*
- ☆ The Council of *Codex Alimentarius Europaeus* adopted a resolution proposing that its work on foods standards be taken over by FAO and WHO
- ☆ The ECE, OECD, the Council of the *Codex Alimentarius Europaeus* established the Codex Alimentarius
- ☆ The FAO Conference decided to establish a Codex Alimentarius Commission.
- ☆ The joint FAO/WHO Food Standards Conference requested the Codex Alimentarius Commission to implement a joint FAO/WHO food standards programme

Need for Full Supply Chain Integration

- ☆ Decision Support
- ☆ Electronic Customer Relationship Management

EUREPGAP (European Retail Parties Good Agricultural Practices) certification procedure

1. Trace ability
2. Record Keeping
3. Varieties and Rootstocks
4. Site Histories and Site Management
5. Soil and Substrate Management
6. Fertilizer Use
7. Irrigation
8. Crop Protection
9. Harvesting
10. Post-Harvesting Treatments
11. Waste and Pollution Management, Recycling and Re-use

12. Worker Health, Safety and Welfare
13. Environmental Issues
14. Complaint Form
15. Internal Audits
16. National and International Legislation

Key Areas for Action and Intervention, (A + 6Cs)

A stands for AWARENESS and ADVOCACY

- ☆ The First C stands for CONNECTIVITY
- ☆ The Second C stands for CAPACITY
- ☆ The Third C calls for CONTENT
- ☆ The Fourth C calls for CREATIVITY
- ☆ The Fifth C denotes the need for COMMUNCIATIONS and NETWORKING
- ☆ The Final C calls for CASH: organizing institutional arrangement for resources

Entrepreneurship Development the key to Economic Development of a Country

Development of Entrepreneurs means Inculcating

- ☆ Entrepreneurial traits in persons
- ☆ Imparting requisite knowledge, development the technical managerial
- ☆ Financial and marketing skills
- ☆ Building the entrepreneurial attitudes

Need for Target Group Oriented Approach

Consisting of young entrepreneur farmers/Farmers practicing commercial agriculture/Specific group of small/marginal farmers

Application of MS ISO 9002 quality system

The key areas are quality, business, financial and risk management

Key to Entrepreneurship Development

Feasibility Checklist

A feasibility checklist is a tool that is used to evaluate the potential success of a new or existing business opportunity

Business Plan

A business plan is vital in determining the credibility of a new business opportunity. It can be used as a management tool and may be required if financing is needed. It also demonstrates to prospective investors a well defined course of action.

The Existing Extension Services

- ☆ Largely in the public sector
- ☆ Other operators (private, NGO, etc.) whenever they exists, remain at the periphery without clear policy enunciation or institutional support.

The upward looking structure of the present extension system generates :

- ☆ Uniformity rather than specificity,
- ☆ Focuses on form rather than substance,
- ☆ Leaves functionaries with low knowledge and low incentives to contribute and perform meaningfully

The key question which need addressal:

- ☆ What changes, in the role of public extension
- ☆ What policy enunciation is necessary to develop other viable and sustainable sources of extension to farmers.

Farmer centered, demand driven and responsive extension services need very clear task and objective clarity.

It also Needs Flexibility of Operation

- ☆ Planning, programming and funding.

Creating Plurality of Information Providers

- ☆ To improve outreach and quality
- ☆ reduce budgetary burden on public exchequer is a desirable objective.

Public Extension Service would Need to Decide

- ☆ Its portfolio of activities based on competitive advantage,
- ☆ Cost effectiveness and initiatives by other actors.
- ☆ Their role may be viewed as promotional and catalytic to establish professionally managed support systems.

We Need

- ☆ **Moving away from individual contact farmer approach to homogenous groups:**
  - based on social, cultural, geographical, gender and commodity,
  - their identification, cohesion, group formation, sustenance and their nurturing.
- ☆ **Collaborative professional activities could be :**
  - technology assessment, technology promotion and evaluation,
  - joint decision making on release of recommendations,
  - joint training.

☆ **Resources interdependence**

- seeking of resources for specific activities,
- release of resources linked to specific work
- Decentralization and dispersal of adaptive research, flexible funding at state/district level.

Linkage

☆ Any collaborative effort between the public and public sectors in which each sector contributes to the planning, resources and activities needed to accomplish a mutual objective.

☆ A relatively new and diverse body of theoretical and empirical literature suggest.

Public-Public partnerships are a constructive means of enhancing the production of goods, services and technologies that would not otherwise produced by either sector acting alone.

In the institutional economics literature, public-private partnerships are viewed as a governance strategy designed to minimize transactions costs, or the costs associated with forming and sustaining relationships – contracting, coordinating, and enforcing a relationship – between actors engaged in production of good or service.

Perspectives of Agriculture Extension Research

Agricultural extension with the initial emphasis on application of scientific research and new knowledge to agricultural practices through farmer education now encompasses a wider range of communication and learning activities organized for rural people by professionals from different disciplines. The term *extension* was first used to describe adult education programmes in England in the second half of the 19th century; these programmes helped to expand - or extend - the work of universities beyond the campus and into the neighbouring community. The term was later adopted in the United States of America, while in Britain it was replaced with 'advisory service' in the 20th century. A number of other terms are used in different parts of the world to describe the same or a similar concept as 'lighting the path' (Dutch), 'advisory work' (German), 'simplification' (French), 'improving skills' (Spanish), 'to promote' (Thai, Lao), 'to promote and extend' (Persian) (Wikipedia)'

There are various definitions of agricultural extension explaining its various perspectives. The perspectives varies widely but essentially can be synthesized as a system of out of school education for application of science for increasing efficiency in agriculture by assisting farm people for capacity building and change in behaviour in resolving problematic situations.

The examples of some of the definitions and its emphasis are given below:

Definitions	*Emphasis*
1949: The central task of extension is to help rural families help themselves by applying science, whether physical or social, to the daily routines of farming, homemaking, and family and community living (Brunner, E. and Hsin Pao Yang,E, 1949)[2]	Application of science
1965: Agricultural extension has been described as a system of out-of-school education for rural people (Saville, A.M.,1965)[3]	System of out-of-school education
1966: Extension personnel have the task of bringing scientific knowledge to farm families in the farms and homes. The object of the task is to improve the efficiency of agriculture. (Bradfield, D.J., 1966)[4]	Increasing efficiency of agriculture
1973: Extension is a service or system which assists farm people, through educational procedures, in improving farming methods and techniques, increasing production efficiency and income, bettering their levels of living and lifting social and educational standards. (Maunder, A., 1973)[5].	System of assisting farm people
1974: Extension involves the conscious use of communication of information to help people form sound opinions and make good decisions (Van den Ban, A.,1974)[6]	ICT for sound opinions and make good decisions
1982: Agricultural Extension: Assistance to farmers to help them identify and analyse their production problems and become aware of the opportunities for improvement. (Adams, M. (1982)[7]	Capacity building
1988: Extension is a professional communication intervention deployed by an institution to induce change in voluntary behaviours with a presumed public or collective utility. (Roling, N., 1988)[8]	Change in behaviour through communication intervention
1997: Extension [is] the organized exchange of information and the purposive transfer of skills. (Nagel, U. J.,1997)[9]	Organizing exchange of information
1999: The essence of agricultural extvension is to facilitate interplay and nurture synergies within a total information system involving agricultural research, agricultural education and a vast complex of information-providing businesses. (Neuchatel Group, 1999) [10]	Facilitate interplay and nurture synergies within a total information system
2004: Extension [is] a series of embedded communicative interventions that are meant, among others, to develop and/or induce innovations which supposedly help to resolve (usually multi-actor) problematic situations. (Leeuwis, C. and Van den Ban, A.) [11]	To resolve (usually multi-actor) problematic situations through a series of embedded communicative interventions to develop and/or induce innovations

Agricultural Extension in Asia

The agricultural extension in Asia can be broadly classified into the following four generations as adapted from NAFES (2005).[12]

Approach	*Emphasis*
Colonial agriculture	☆ Focus on export crops ☆ Technical advice to plantation managers and large landowners ☆ Rare assistance to small farmers, except in times of crisis
Commodity-based extension services	☆ Production targets with various schemes to meet the needs of small farmersmeet
Unified extension through	☆ Single national service
Training and Visit system (T&V)	☆ Regular messages to groups of farmers for promoting adoption of technologies
Growing concern for sustainability and equity	☆ Participatory methods

The present thinking is that the agricultural extension needs to be reinvented as a professional practice (Leeuwis, C. and Van den Ban).[13]

Research Areas

In the process agricultural research, in all the other disciplines, basic or fundamental, strategic, applied and adaptive stages of research are carried out. The basic research provides basic understanding which is a pre-requisite knowledge to be incorporated for strategic research. Further research is taken up with the application of such knowledge for development of system, situation and commodity-specific modules and capsule for agricultural extension education and services. The aim of adaptive research is to assess and refine by making selection of the modules and capsule to be precisely effective in order to facilitate such technological change in a given social system. For example, working out adult learning principles is basic research, incorporating these principles in Experiential Learning Cycle is strategic research, designing a training session based on Experiential Learning Cycle is applied research and evaluation of the training session in terms of its relevancy and effectiveness in different agricultural topics is adaptive research. Similarly, identifying types of capability of learning is basic research, arrangement of such capabilities in hierarchical fashion is a strategic research (Bloom's Taxonomy), and designing different training sessions based on the needed capability of learning objective is applied research.

A list of examples of probable research areas is given below:

Sl.No.	*Research Areas*	*Type of Research*
1.	Determine the new and emerging skills needed by the farmers for improved sustainability in different farming systems	Strategic
2.	Characterization of complexity, diversity and risk proneness of agriculture in different agro-ecological regions.	Basic
3.	Identify instructional materials needs to different complex, diverse and risk prone agriculture.	Strategic

Contd.

Sl.No.	Research Areas	Type of Research
4.	Determine models for sequencing competency development for quality production programmes	Basic
5.	Identify standards for providing extension service	Basic
6.	Evaluate factor productivity of agricultural education programmes	Applied
7.	Establishing the relationship between extension education, extension service and increased productivity.	Basic

Conclusion

In the context of present scenario of agriculture, there is need for a paradigm shift in research in agricultural extension. Extension research needs to be more broad-based and holistic both in content and scope, much beyond determining impact of socio-economic factors for transfer of technology of post mortem in nature. The scientists engaged in the task of research in agricultural extension need to develop a more proactive approach in generating knowledge/information of concept, approach and methodology in facilitating relevancy and effectiveness of extension education programmes and services.

There is need for research in micro level institutional restructuring, management of extension programme, models of research-extension-civil society linkage, capacity building of farmers, models of empowerment, and mainstreaming gender implication in agriculture and overall sustainability of agricultural extension as a discipline. Generation of concept and methodology and its application will have to be location specific with appropriate optimization and input-output/outcome/impact to agro-ecological or social circumstances for its use by the planners, policy makers and all those engaged in the task of promotion of productive, stable, sustainable, equitable and profitable farming.

Chapter 20

Technological Innovation and Extension Approach for Island Ecosystem

R.C. Srivastava

Andaman and Nicobar Islands are our eastern most frontier and have a very strategic role in security of our vital economic installation and natural resources. Further they are warehouse of vast biodiversity. Even though the agriculture was known to have been started in the Islands as early as 1857 with the penal settlement, it had never been a significant vocation either during British occupation or even after independence until early sixties. At the most, the land was utilised for grazing or for cultivation of plantation crops and vegetables. The agricultural activities as a subsistence living was started by the Mopla exiles and ex-convicts in a few villages of South Andaman. Soon after Independence, the burden of bearing the large influx of refugees erstwhile East Bengal prompted the Indian Government to settle some of them in these Islands by allotting mostly 2 ha each of paddy and hilly lands. The settlers, forming nearly 90 per cent of the total population, include ex-convicts of the British penal colony and their descendents, refugees from Bangladesh, repatriates from Sri Lanka, ex-servicemen, migrants from various parts of the sub-continent, labours who were brought in mainly from Bihar to work with Andaman Public Work Department, a large number of govt. and public sector personnel, traders and fishermen families from mainland. The Andaman tribes are primarily hunters, food gatherers while Nicobar tribes are mainly horticulturists and herders. It is natural that there are no traditional agriculturists and the main agriculture is practiced by the displaced persons who are settled in these islands especially from Bangladesh and those from Kerala. Most of the agricultural activities remain with the established settlers hailing from erstwhile east Bengal. The present population in the islands is about 4 lakhs. The total land area is 8249 sq. km, out of which lush green tropical forests cover about 86 per cent. Out of the total 45612 ha cultivable land, rice covers 7583 ha area, coconut and arecanut covers 20891 ha. and 2251 ha., respectively. Fruits, spices, cashew, rubber and oil palm are also important cash crops of these islands.

Thus, the agricultural development in these islands can not be viewed in isolation. Thus the real challenge to the future island agriculture is to meet the requirement of local population and tourists at a reasonable cost.

CARIs Role in Furthering the Cause of Agriculture Development

Indian Council of Agricultural Research (ICAR) established Central Agricultural Research Institute (CARI) at Port Blair on June 23rd, 1978 by merging different regional research stations of ICAR institutes located in islands. Situated in the rural background the institute has been instrumental in supporting the island agriculture through technological intervention leading to livelihood opportunities for the people both during pre and post tsunami situation.

During the last thirty years, the institute has been able to achieve its mandate of reducing the gap between requirement and local production of different commodities. The contribution of local production to the requirement ha risen during last thirty years (1980-2008) from 25 per cent to 50 per cent for rice, tropical fruits from 40 per cent to 100 per cent, flowers from insignificant to 25 per cent, vegetables from 20 per cent to 80 per cent except potato and onion which is zero, milk from 20 per cent to 70 per cent, eggs from 25 per cent to 90 per cent, chicken from 10 per cent to 100 per cent, inland fishes from 10 per cent to 90 per cent and pulses from insignificant to 20 per cent.

Table 20.1: Contribution of Local Production to Requirement

(As there is no increase in land area, the increase has come from technology adoption)

Commodity	*1980*	*2006*
Rice	25 per cent	50 per cent
Tropical fruits	40 per cent	100 per cent
Flowers	Insignificant	25 per cent
Vegetables	20 per cent	80 per cent
Milk	20 per cent	70 per cent
Egg and Chicken	10-20 per cent	100 per cent
Fresh water fishes	Insignificant	90-100 per cent

Source: CARI Perspective Plan – Vision – 2025.

Since there has been no significant increase in the land area under cultivation, the increase has mainly come from technological inputs, which has been result of research and extension efforts of this institute. This effort has to continue with more vigour to meet the new challenges arising out of increasing population and tourist traffic. The islands have a fragile ecosystem with tropical rain forests which forbids any horizontal expansion of cultivated area.

Redefining the Extension Approach for Tropical Island Conditions

In view of the socio- economic –cultural background of the farmers, the rising aspiration level of the farm youth, fragility of island ecosystem which do not allow any horizontal expansion of land, the traditional extension approach will not hold good for making agriculture competitive and enable it to fulfil the role in island economy. In view of this, the institute has come forward with innovative extension approach which include *viz.* Integrated Development of the Villages, Integrated Farming System and Convergence of technologies as Micro Enterprises for improving efficiency, profitability and finally ensuring decent livelihood.

(1) Integrated Development of Village

The TOT programmes have to be taken up as a group approach rather than an individual farm family approach. For the successful implementation of the village development programmes, CARI which is the only multi disciplinary Institute has functional links with the State Development Departments like Agriculture, Animal Husbandry and Veterinary Services, Fisheries, Rural Development, Health services, Social Welfare Advisory Board, Nationalized Banks, NABARD, PRIs and the NGO's. This has ensured in facilitating a holistic approach within research – extension – farmers – marketing *i.e.* out of the four pillars, CARI acts as one of the pillar which provide technological backstopping in Agriculture and allied fields, the line departments (second pillar) provides the input required for the technology, the NGOs, banks, the third pillar provides the marketing linkage and PRIs (Fourth pillar) provides a congenial environment and better governance for bringing all round development of the village and also to reach the unreached in the far-flanged islands of these territory. Such approach has been found successful as a role model for its duplication by the stakeholders.

(2) Convergence of Technologies as Micro Business Module

National Sample Survey envisages that about 40 per cent of the farmers are ready to leave agriculture if an alternate livelihood option is available. Similarly 60 per cent of practicing farmers are above 50 years, which means that the youth are not viewing farming as an occupation. The situation in the islands is even more peculiar wherein the first generations of settlers who were either freedom fighters/convicts or displaced persons from Bangladesh/Burma/Sri Lanka took up farming as it was the only source of livelihood. However with globalization and information explosion, the aspiration level of today's youth has risen. This young population does not find quality of employment available in traditional agriculture compatible to their aspirations to lure them towards farming. Thus our agricultural production system has to be reoriented by appropriate intervention for reducing drudgery, value addition, increased productivity and improved profits to match his aspiration.

To retain farm youth in agriculture, an innovative approach of translating fourteen technologies in Agricultural and Allied fields as **Micro Business Modules (MBM)** for providing **decent livelihood** options to the youths of the island has been developed in association with NABARD. The details of the technology is available on our website http://cari.res.in. The investment requirement varies from as low as Rs 20000/- to a maximum of Rs 2.5 lakhs suiting every body's budget and risk taking capacity. (Srivastava and Ahmed 2008).

A brief information on the technology, area required, minimum investment and net return befitting to different socio- economic situation is given in the table below:

Table 20.2: Technology, Area Required, Investment and Net Return from Livelihood Options

Technology	*Area/ Land Needed (Unit)*	*Minimum Investment*		**Net Return*	
		Duration (in Months)	*Amount (In Rs.)*	*Duration (Month onwards)*	*Amount (In Rs.)*
Integrated Farming System (IFS)	2000 m^2	16	1,40,000	40	1,53,565
Broad Bed and Furrow (BBF) System	2000 m^2	12	40,500	24	31,825
Raised bed technology	1153m^2	12	18,981	12	33,179
Oyster Mushroom cultivation	46m^2	12	59,000	12	21,000
Black pepper cultivation on hedgerows with *Gliricidia* standards	10,000 m^2	12	36,200	60	2,90,600
Floriculture (crossandra, marigold Land tube rose)	10,000 m^2	12	64,500	12	98,000
Protected cultivation of High value vegetables	75 m^2	12	11,500	12	58,500
Goat Farming	4500 m^2	25	2,70,200	32	2,37,340
Pig Farming	600 m^2	16	2,35,350	28	1,49,820
Quail Farming	1000 m^2	12	2,55,000	12	99,625
Mud Crab Culture	1,000 m^2	6	2,600	6	2,080
Marine Ornamental Fishes (Damsels)	2000m^2	12	1,34,000	12	1,06,000
Small scale seed production and nursery raising of Indian major carps	2000m^2	12	2,57,000	24	1,63,000
Vermicompost Production	Unit capacity 10t/ha.	12	28,000	12	62,000

* Net return is after deducting the fixed cost

(3) Integrated Farming System (IFS) for Sustainable Productivity of Island Ecosystem

Agricultural transformations have to occur at the level of small holding farmers so that these complex-farming systems can be made more productive and efficient in the use of resources. Even in these islands there is an alarming increase in population and the production of various commodities are falling short of demand. The land area, specially cropped area cannot be substantially increased. Multiple land use through integration of crops, livestock and aquaculture can give the best and optimum production from unit land area. The concept of faming system, which brings together all these components to make it wholesome system, is a paramount importance as it ensures for optimum land use, maximum return, soil conservation, build up of soil fertility, better use of production resources, recycling of wastes freedom from pollution, providing year round employment for farming family, risk management, reduction at the cost of production and supply of essential commodities throughout the year.

Over here we are giving you the characterization of the existing micro farming situations, components of the IFS, recommended models and profitability in terms of net return (Ravisankar *et al.*, 2007 and Srivastava *et al.*, 2009).

Characterization of Farming System

Agro-ecosystem analysis of the farming systems in these islands revealed four distinct micro farming situations (MFS) in the farmers' field. They are

MFS I : Hilly

MFS II : Slopping hilly upland

MFS III : Medium upland valley

MFS IV : Low lying valley

Hills (MFS 1)

Coconut and arecanut is the main crop with fruit trees like jackfruit, guava, sapota as intercrops in the backyard of homesteads: Topography is sloppy, soils were gravel and coarse sand, soil depth was 30-45 cm only with underlying stones and boulders. Major problems are soil erosion. Ponds are present in the lower reaches of the slope. Cattle rearing and backyard poultry are the allied enterprises in this situation.

Slopping Hilly Upland (MFS II)

Arecanut is the main plantation crop, banana, pineapple are grown as fruit crop. Majority of the cash crops like vegetables, flowers, sugarcane, are cultivated on this land, topography is slopping 15-25 per cent, soils are coarse sand, gravel, soil depth is 50-70 cm., soil erosion is the major problem. Unmanaged fishponds are present. Backyard poultry, cattle and goat rearing is common (Ravisankar and Pramanik, 2007).

Medium Upland Valley (MFS III)

Paddy, vegetables and pulses are the main crop, topography is less slopping. This land mainly are used for HYV Vegetables cultivation as cash crops, soil are sandy loam and silty due to colluvial deposition. Soil depth is more than 1.5 m. Fishery, backyard poultry, duck and goat rearing is undertaken.

Low Lying Valley (MFS IV)

Long duration paddy is the main crop. Fields are adjacent to seacoast. Soil depth is more; soil is clay to clay- loam. Water logging during monsoon and post monsoon dryness is the main problem. Backyard poultry and cattle rearing are common. Some farmers also take up fishery.

Components of IFS

All the models recommended are for one ha area except medium upland valley micro farming situation where in the area available for cultivation is only around the pond. Hence, the area for this micro farming situation was taken as 0.15 ha.

I. Hills (1 ha)

Crop (plantation crop) + cattle + backyard poultry: In this MFS cropping is integrated with cattle rearing and backyard poultry.

Cropping Sequence

Coconut (0.01)+ sapota (0.05)+ banana(0.05)	: 0.20 ha
Arecanut + black pepper+ fodder	: 0.50 ha
Sapota + banana (0.03 ha)	: 0.10 ha
Banana	: 0.10 ha
Livestock and Poultry	: 0.10 ha

II. Slopping Hilly Upland (1 ha)

Crop (paddy, sugarcane, vegetable, floriculture and plantation) + cattle + fish + poultry: In this MFS paddy, sugarcane and vegetables are main crops. This rotation provides maximum soil cover index and better nutrient management of the soil. Fodder and spice like black pepper is integrated with plantation crop. Cattle are reared on crop waste and fodder and the manure obtained is composted and applied to the field. Fish cum duck cum poultry is another important integrated system in this MFS. For a pond area of 0.036 ha, 20 birds + 10 ducks with 200 Indian major carp fingerlings are recommended. Floriculture is introduced in this MFS and has proved successful in enhancing the net return.

Cropping Sequence of Model 1 and 2

Model 1

Floriculture – leafy vegetable – okra	0.10 ha
Paddy – sugarcane (Chewing canes)	0.22 ha
Ginger – leafy vegetable – leafy vegetable	0.16 ha
Paddy – vegetable – vegetable	0.19 ha
Coconut + fodder	0.08 ha
Arecanut + black pepper	0.05 ha
Coconut + crossandra (.025)+ marigold (.025)	0.05 ha
Coconut + ginger	0.05 ha
Fish cum poultry cum duck	0.036 ha
Livestock and fodder	0.064 ha

Model 2

Paddy – vegetable (okra)– floriculture	0.0 5ha
Paddy – vegetable (cowpea)– floriculture	0.0 5ha
Paddy – Paddy – vegetable (okra/brinjal)	0.10 ha
Paddy – Sugarcane (Chewing canes)	0.20 ha
Paddy – vegetable (S.G, B.G, M and A, each 0.05ha)	0.20 ha
Paddy – Maize	0.02 ha
Sugarcane – Sesame	0.01 ha
Ginger – leafy vegetables – ginger	0.16 ha
Coconut + fodder	0.02 ha
Arecanut + black pepper + Crossandra	0.02 ha
Coconut + crossandra	0.07 ha
*S.G = Snake gourd, B.G + Bitter gourd,	
M = Marigold and A = Amaranthus	
Fish cum poultry	0.036 ha
Livestock and fodder	0.064 ha

III. Medium Upland Valley (0.15 ha)

Crop (vegetable, plantation) + goat + backyard poultry + fish + poultry: In this MFS plantation is integrated with fodder for the animals. Vegetables are the main crop in this MFS. Goats are reared in a semi intensive type of rearing, they are let free for grazing and in the night shelter they are provided fodder and crop residue. The manure collected is applied in the field as well as in the fish pond. Backyard poultry rearing generates extra income with low input cost. Fish cum duck cum poultry integration is another promising subsystem in the IFS.

Cropping Sequence of MFS – III

Vegetable (okra)	0.02 ha
Vegetable (cowpea)	0.01 ha
Vegetable (bottle gourd)	0.01 ha
Livestock and fodder	0.05 ha
Fish cum poultry cum duck	0.06 ha

IV. Low-lying Valley (1 ha)

IFS: Crop (paddy, vegetable and plantation) + cattle + backyard poultry + fish + poultry: Paddy and vegetables are main crops in this MFS. The vegetable production is low in this MFS due to water logging in heavy downpour. Bullocks are reared for ploughing in the field and cows are reared for milk. The manure obtained is used in the form of compost. Poultry is integrated with fishery, which reduces the fish feed cost.

Cropping Sequence

Paddy (long duration)	0.20 ha
Paddy (Short duration) – vegetables*	0.61 ha
*Vegetables (okra - 0.12 ha, bottle gourd - 0.14 ha, pumpkin – 0.10 ha, ash gourd – 0.10 ha, cowpea – 0.15 ha)	
Coconut + orange + fodder	0.02 ha
Areca nut + black pepper	0.05 ha
Vegetables on pond dike (bottle gourd and ridge gourd)	0.02 ha
Livestock and fodder	0.064 ha
3 cows + 2 bullocks	0.004 ha
Fodder	0.06 ha
Fish cum poultry cum duck	0.036 ha

Broad Bed and Furrow (BBF) System

Since vegetables are fetching higher price during monsoon season, broad bed and furrow system developed by Central Agricultural Research Institute, Port Blair for growing vegetables in the bed and paddy + fish in furrows gives better option for farmers for realizing higher income from unit land area (Ravisankar *et al.*, 2003, 2009 and Srivastava *et al.*, 2008).

Profitability in Terms of Net Return

Hilly

Net return obtained from cropping was Rs 3,26,408, livestock and poultry was Rs 64,300/ha.

Slopping Hilly Upland

Model 1 Net return obtained from cropping was Rs 1,53,989 livestock and poultry was Rs 77,176 and poultry + duck + fish was Rs 12,465.

Model 2 Net return obtained from cropping was Rs 2,94,349 livestock and poultry was Rs 84,946 and poultry + duck + fish was Rs 14,872.

Medium Upland Valley

Net return per ha. area system obtained from cropping was Rs 9,297 livestock was Rs 72,306 and poultry was Rs. 2,630 and poultry + duck + fish was Rs 24,957.

Low Lying Valley

Net return from IFS obtained from cropping was Rs 57,760 livestock was Rs 90,893, backyard poultry was Rs 11,000 and poultry + duck + fish was Rs 14,872.

In BBF system, among the cropping system, okra-amaranthus-okra recorded maximum net return of Rs. 1,06,134 from 4000 m^2 area of beds in one ha, Rs 10840 was obtained from furrows. Thus, from one ha of BBF, through adoption of okra-amaranthus- okra in beds, brinjal+moringa+banana in border areas and rice +azolla +fish in furrows, Rs117532/ha was obtained, which is eleven times higher than the normal cropping of paddy followed by pulses/fallow.

Per cent Contribution of different Enterprises

Hilly

Contribution from cropping, livestock and poultry was 84, 13 and 3 per cent respectively.

Slopping Hilly Upland

Model 1 Contribution from cropping, cattle, goat, backyard poultry and poultry+duck+fish was 69, 13, 10, 2 and 6 per cent respectively.

Model 2 Contribution from cropping, cattle, goat, backyard poultry and poultry+duck+fish was 74, 16, 5, 1 and 4 per cent respectively.

Medium Upland Valley

Contribution from cropping, cattle, backyard poultry and fish cum poultry cum duck integration was 9, 66, 2 and 23 per cent respectively.

Low Lying Valley

Contribution from cropping, livestock and fishery was 49, 50 and 1 per cent respectively.

Water Based IFS

After Tsunami 2004, many farm ponds were constructed under Rajiv Gandhi Rehabilitation programme. The present estimate shows that around 1700 ponds are available in the islands which are not being utilized for multiple purposes. So these ponds can be developed as a full fledged production system depending upon their water characterstics (fresh and brackish) by integrating crop, animal and fish components (Srivastava *et.al*. 2008 and Annual Report 2006-07, CARI, Port Blair).

Freshwater Pond Based IFS

The embankment can be effectively used to grow crops like coconut, arecanut, banana, papaya along with vegetables like pumpkin, bitter gourd, etc. Fodder scarcity being a major constraint in the milk production system in the islands can also be introduced in both inner and outer slopes which also protected the ponds against erosion. Poultry can be introduced as additional income source over the ponds, whose droppings can be utilised as the manure source for crop plants. The details of income generation is presented in Figure 20.1

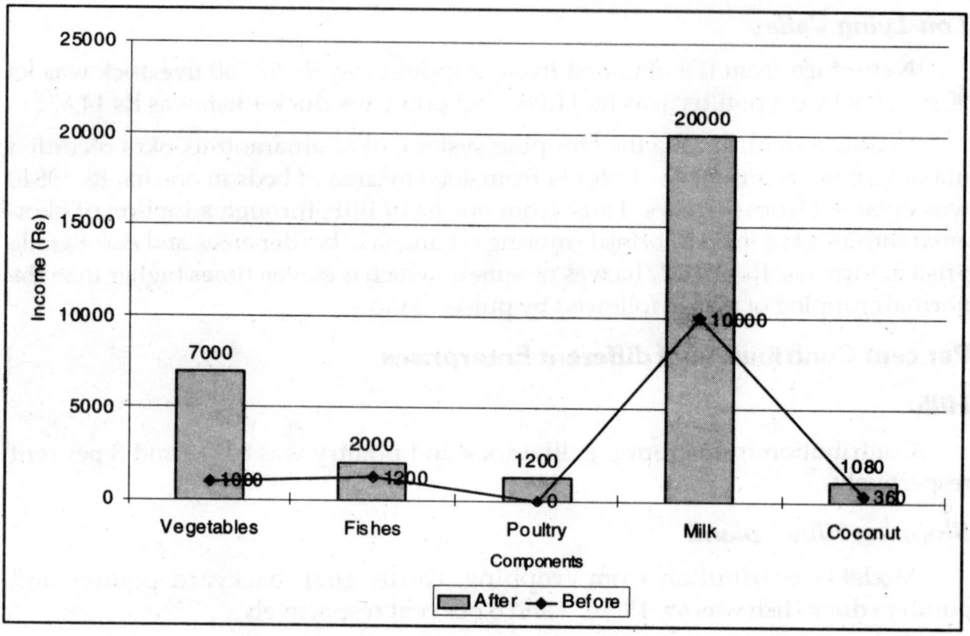

Figure 20.1: Pre Intervention and Post Intervention Income from different Components of Freshwater Pond Based IFS.

It is very clear from the Figure 20.1 that the income from the scarce land area increased significantly attributing to the overall improvement in the livelihood of the farmer. The income of the farmer increased from Rs. 12860/- to Rs.31280/- in one year from a pond of 1500m^2 area.

Brackishwater Pond Based IFS

Along the dyke crops can be taken in beds of size 1x 5 x 0.1 m^2 such as spinach and pumpkin. Plantation and fruits crops of coconut, Banana, Morinda can also be integrated. The identified fodder grown naturally in the inner and outer slopes of the embankments of brackish water pond is a variety of buffalo grass whose palatability was good and can be recommended for feeding to the cattle and goats.

To sum up few initiatives taken up by the institute in the recent past is also been highlighted which has given way to better probabilities for the innovative stakeholders who are always in search of new avenues giving better returns namely:

- ☆ Introduction of table purpose rabi groundnut along with seed production facility during kharif.
- ☆ Evaluation of long duration paddy varieties (160-170 days) to take advantage of long rainy season.
- ☆ Rain water management technology for water resource development at hill top and valleys.
- ☆ Fodder cultivation on hill slopes through grasses and fodder trees.

- ☆ Reduction in use of pesticide by using IPM techniques.
- ☆ Intensification of work on flowers, banana, papaya, pineapple and off season mango.
- ☆ Black pepper village with black pepper on *Gliricidia* as standards and value addition through proper cleaning, grading and packaging.
- ☆ Integrated farming system in fresh and brackish water areas.
- ☆ Induced breeding of cat fish and development of hatchery.
- ☆ Technology of captive breeding of Damsel ornamental fishes.
- ☆ Introduction of buffalos and management practices for indigenous cattle.
- ☆ Low cost and alternate feed for poultry.
- ☆ Protected cultivation for high value crops and their value addition.
- ☆ Identification of technology package for farmers of different socio-economic background.
- ☆ Demonstration of four identified technologies namely Broad Bed and Furrow System, Integrated Farming System, Tank cum Well System and Micro Irrigation System in farmers field in participatory mode.
- ☆ Development of comprehensive technology package for *Morinda citrifolia* (Noni) to make it a commercial plantation crop of islands.
- ☆ Cage culture in bay/creeks.

Conclusion

Based on the study, it can be concluded that, the Integrated development of the villages for holistic development, plantation based cropping system along with animal component in hilly areas while floriculture and vegetable based integrated farming system for slopping hilly upland areas should be promoted in order to achieve higher returns and sustainable use of farm resources. In case of valley areas, where in two kinds of situations exists *i.e.* for medium upland valley and low lying valley, it is suggested that, pond based integrated farming system involving fish + poultry+duckery+vegetables + cattle for medium upland valley areas while, either the Broad bed and furrow based system or animal based integrated farming system should be promoted for realizing higher unit area land productivity and sustainable income for the farmers and finally the micro- enterprise development to make them an Agri-pruner for ensuring decent livelihood. In order to achieve the full production potential of water bodies created, fresh and brackish water based IFS should be promoted.

References

R.C. Srivastava, 2009. CARI in service of Island – Achievement in 2008 and projections for 2009, page 3, Jan. 1, *The Echo of India, Port Blair.*

N. Ravisankar, S.K. Zamir Ahmed, S.K. Ambast and R.C. Srivastava, 2009. Crop Diversification through Broad Bed and Furrow (BBF) System, p-3, February 4th, *The Echo of India, Port Blair.*

Srivastava, R.C and S.K. Zamir Ahmed. 2008. Micro Business module in Agriculture and allied fields for livelihood options in A and N islands, p-74, *Central Agricultural Research Institute, Port Blair*

R.C. Srivastava, S. Dam Roy and S.K. Ambast, 2008. CARI : A journey of 30 years in service of Island farmers, p-120, *CARI, Port Blair.*

R.C. Srivastava, T. Damodaran, N. Ravisankar, S. Jeyakumar, S.K. Zamir Ahmed, M. Balakrishnan and C.S. Chaturvedi, 2008. Integrated Rehabilitation of Tsunami affected farmers through technological intervention in Andaman and Nicobar Islands, p – 72, *CARI, Port Blair.*

R.C. Srivastava, N. Ravisankar, S,K, Ambast and S.K. Zamir Ahmed, 2008. AI Swaminathan adopts Integrated Farming System in the Island, page 3, December, 17, 2008, *The Echo of India, Port Blair.*

Ravisankar, N., S.K. Ambast and R.C. Srivastava. 2008. Crop diversification through broad bed and furrow system in coastal regions, p-180, *Central Agricultural Research Institute, Port Blair*

CARI Perspective Plan – Vision – 2025, 2007 p-58, *CARI, Port Blair.*

Ravisankar, N., S. C. Pramanik, R. B. Rai, Shakila Nawaz, Tapan Kr. Biswas and Nabisat Bibi. 2007. Study on Integrated Farming System in hilly upland areas of Bay Islands, *Indian Journal of Agronomy* **52** (1):7-10

Ravisankar, N. and S. C. Pramanik, 2007. Integrated farming system mode for hilly uplands of Bay Islands, *Indian Farming,* **57**(8):11-16.

Annual Report 2006-07, CARI, Port Blair p-192.

Ravisankar, N., R. Raja, R. Dinesh and S. Ghoshal Chaudhuri, 2003. Crop diversification through broad-bed and furrow land manipulation in coastal paddy lands, *Natn. Symp. on Crop production under changing environment,* 27-29th November 2003, Bidhan Chandra Krishi Viswavidyalaya, Kalyani, Nadia, West Bengal, India. P.52.

Chapter 21

Redesigning Extension Strategies and Policies for Sustainable Growth

Daulat Singh and Lakhan Singh

Concept of Frontline Demonstration

Frontline demonstrations have served only frontline farmers having adequate resources, extension contacts and those who are front in the farming community. The majority of small and marginal farmers disadvantaged of resources, extension contacts and societal status were left behind for a quite long period. This scenario can only be changed to the class of disadvantaged farming community through directly exposing them to the new technologies in their tiny plots with resource inadequacies.

The suggested approach will also accelerate the process of exchange of seed from farmer to farmer as large number of farmers will have the innovations to share with other left out farmers. The exchange of seed can be facilitated by motivating farmers to return a small quantity of seed to the next farmer, which who will be willing to do. The exchange may also be promoted through harvested produce to labourers. Even self help group can create a small seed bank for their members to use it. Quantity of exchange of seed between individual farmers would be so small that even it can cover exchange of seed in between and given without much hesitation. In real terms, the concept of SHG needs to be oriented with more group activities and interactive themselves.

The process of implementation of *'Mini Seed Demonstration cum Exchange Programme'* will include analysis of the existing local varieties, assessing their purity, germination and real value, comparing it with recommended varieties. The size of demonstration will be the common size for the marginal farmers. This mini demonstration of seed cum exchange has been initiated in the NAIP project of ICAR Concept-3 *'Improvement of rural livelihood security of rural people living in disadvantaged districts through diversification in agriculture'* in two districts (Fatehpur and Hardoi) out of 15 most disadvantaged districts of Uttar Pradesh.

Convergence of Schemes

There is large number of schemes funded and directed by the Government of India and State Governments for the benefits of rural communities for their improvement in livelihood. But unfortunately, all these schemes water whirled and not settling down to the groups for which they are basically developed. Some schemes are running parallel. They do not club and give the benefits to the target.

There is no convergence of schemes either at block level or SHG level being organized by all the organizations. Number of SHGs floating with number of benefits of designed schemes to all the members of groups. This anomaly to some extent can be reached if the development schemes approved at village level, can be further converged among the members of SHGs. It is gigantic task but beginning can be made in the projects like NAIP-Concept 3 primarily developed and directed for improvement in livelihood security of farming community through diversified approach.

Modeling of SHGs

Most of the SHGs stop functioning as soon as financial credit is disbursed along with subsidies. The observations suggest the recovery of credit is dismal. There is a need to add more activities in the group rather than adding liberal funds to them. This liberal approach has disturbed and changed the attitude of several beneficiaries.

The SHGs developed on the pattern of production-micro saving-inter-loaning-micro financing-micro enterprise and micro insurance have proved better in creating positive attitude and better understanding of the group activities and their benefits to the group as whole. Only those SHGs continue to function even after exit of the project where large number of activities was created during project period and those groups in which some of activities continue afterwards. SHGs don not sustain on credit only, they sustain more on number of profitable activities adopted by large number of beneficiaries.

Migration

With the industrial growth and development of infrastructures in the country, the migration has taken a big activity. There has been *rural push* and *urban pull* which has completely changed the social structures in the villages. With the migration of labourers and employable youths to the cities has left the village in a deserted look. Most of the migrants have left their families, children and old parents in the village with no proper responsible guardian under the overall responsibility of the house woman. In actual term, the house wife is the head of the family in the absence of migrants to the cities. The major enterprises now in village development schemes should be focused on making the woman as the head of the family particularly for landless, marginal and small farming community which is the most disadvantaged groups in the village. Under NAIP besides transfer of technology, these organizational aspects of strategies should find a prime place and group formation if not properly organized, nurtured; the sustainability will be a far flunch dream.

Rural Entrepreneurship

Entrepreneurial activities are entering in rural India are very fast. Several companies dealing with consumer goods are establishing their dealers/sub-dealers right in the village to reach the customers first. Several goods which were bought by the villagers in the township are reaching directly in the village. This entrepreneurial change emphasize the need for initiating low cost micro enterprises right in the villages as an important activity of the SHGs and extension strategies should be market-led based on consumers' demand of the rural people. SHGs can be profitably utilized in starting fertilizers, pesticides, seed shops, repair and maintenance facilities of agricultural implants, tools, pumping sets, electric motors, etc. Agricultural sector has lagged behind in creating proper entrepreneurial group activities for meeting their requirements.

The time is opportune for redesigning village extension strategies, policies and approaches for agricultural growth and development especially through involvement of landless, marginal and small farmers in the orbit of their responsibilities. It is possible if we focus the disadvantaged farming community when preparing strategies and policies related to development, transfer and sustenance of agricultural endeavours. Dr. Md Yunis, 2006 Noble Peace Prize awardee and founder of Grameen Bank, Bangladesh once remarked *'Poverty is an artificial creation, is does not belong to human civilization, we can change it and can make people out of it. The only thing we have to redesign our institutions, strategies and policies'*. The time has come that SHGs should be organized on business reasons rather than social obligations for a client. Social development would happen automatically.

Chapter 22

Rebuilding the Confidence of Farmers: Experiences of an Innovative Extension Project

K. Narayana Gowda

According to available data and feed back from several quarters, farmers in general and farm youth in particular are losing confidence in farming. Given the choice more than 40 per cent of farming families wanted to leave farming in Punjab. In a progressive state like Punjab where the farmers are relatively well assured with resources compared to majority of other states in the country, are losing confidence in farming, it is difficult to imagine the mindset of farmers in other parts of the country. The situation is more or less same in many developing as well as few developed countries.

There are many factors that have contributed for disinterest and apathy among farmers. The major factors are indiscriminate increase in cost of production, lack of assured market and price, uneconomical and fragmentation of land holdings, timely non availability of labour, inadequate and timely availability of critical inputs at the easy reach of majority of farmers, decline in soil productivity and depletion of under ground water resources, increasing ecological degradation and ravaging climate change have posed a serious threat to farmers' economy and living conditions. In the process farmers are loosing confidence in the recent days, shifting to non agriculture vocations, migrating to urban areas and in extreme cases they are committing suicide. Youth are in the verge of leaving farming. All those in the helm of affairs today, should seriously think of how to restore the confidence in farmers, particularly farm youth. Otherwise a big vacuum is going to be created in the coming years leading to many unsolvable crisis. Therefore there is need for specific extension strategy to improve the economy and living standards of farming people. Once the farmers' conditions are improved, it will have chain reaction in improving the conditions of other segments of rural people leading to rural prosperity. The experience of Rural Bioresource complex Project implemented by five Centres across the country can be of great opportunity to improve rural economy and living standards.

Department of Bio technology, Government of India has launched five RBCs across the country and one such RBC is implemented by the University of Agricultural Sciences, Bangalore since April 2005. The project envisages to improve the income and standard of living of the intended families in a span of five years. The project is implemented in Tubagere Hobli of Doddaballapura Taluk of Bangalore Rural District covering 8340 families spread over in 75 villages of five panchayats in a contiguous area. Base line information was collected, analysed and documented for the entire families. Majority of farmers were cultivating traditional crops realising 60 per cent of the potential yield. The extension participation of farmers was only 25 per cent. More than 85 per cent of the area is under rainfed with less than 760 mm average annual rainfall. In the last three years, the Project was able to bring about significant changes in productivity, income and employment generation with considerable influence on standard of living.

Special Features of the Project

The achievements were possible due to contribution of six important factors namely; identification of most profitable, sustainable and location specific technologies, timely and dependable information at the easy reach of rural people, providing critical inputs free of cost at the easy access, effective functional linkage, marketing empowerment and commodity based associations. The salient features of these factors are detailed below.

I. Profitable and Sustainable Technologies

Altogether five categories of technologies consisting of 19 interventions were identified and promoted in the project area keeping in view type of land, water availability, marketing opportunity and resource status of stake holders. The lists of these interventions are given below.

Interventions 2007-08

	Nature of Intervention
I	Rainfed
	a) Arable
	1. Improved cultivation of Ragi and redgram
	2. Improved cultivation of Maize (All rounder)
	b) Non-arable
	1. Bio-Fuel Plantations
	2. Dry land Orchard

Contd

Contd.

II	Under Assured irrigation
	a) Hi-Tech Intervention
	1. Improved Rose Cultivation
	2. Improved cultivation of Sweet Corn
	3. Improved Cultivation Baby corn
	b) Other than Hi-Tech Interventions
	1. Improved Ragi Seed Production
	2. Improved Beans Seed Production
	3. Improved cultivation of Hybrid Maize
	4. Improved cultivation of Popcorn
	5. Improved cultivation of Sunflower
	6. Improved cultivation of Mulberry
	c) Subsidiary interventions
	1. Improved Sheep Rearing
	2. Improved Fish Rearing
	3. Improved Poultry Birds Rearing
	4. Agro processing unitValue added Products in Ragi(Ragi Malt)
	5. Water use efficiency
	6. Integrated farming system
	7. Infrastructure for Farmers Associations

II. Information Support System

Inadequate, untimely and undependable information has weaken the mindset of farmers in farming. Project took all the initiatives to provide best information support to stake holders and they are as follows.

1. Eight junior Scientists were recruited to work with 1500 to 2000 families per each junior scientist. They are trained and guided by 30 senior scientists regularly to update their knowledge and skill matching with project mandates. They are provided with motorcycles for which mobility and timely guidance to farmers.
2. Organizing need based training programmes both on campus and off campus on the interventions promoted and other related areas to stake holders.
3. Established one Village Resource Centre (VRC) at Project area and an Expert Centre at GKVK Campus to provide continuous information support to all the stake holders.
4. Other educational activities namely field visits, field days, study tours, melas, Krishi melas, campaigns, exhibitions, and vanamahotsava were organized.

5. Recognitions and awards: Best farmers, Farm women, Farm youth and local leaders are recognised / given awards during field days, Krishimela and other occasions which is the greatest satisfaction, as well as motivating fellow stake holders for achieving excellence.

III. Providing Quality Critical Inputs

The biggest problem many farmers and rural people facing today is inadequate availability of quality critical inputs at the easy reach of stake holders. The project has made special provisions to provide best quality critical inputs to beneficiaries tailored with respective interventions at the appropriate time. Even the seedlings / tissue culture banana seedlings / rose cuttings / fingerlings were provided using the common transport facility. The start of farmers associations / commodity based associations have been instrumental in starting various nurseries in the local area which helped even marginal farmers to take up planting at his convenient time.

IV. Effective Functional Linkage

In all 13 agencies are collaborating with the project representing three research organizations, development departments, commodity Boards and one financial institution.

V. Marketing Empowerment

The major problem faced by all farmers and rural people is the issues relating to marketing of their produces. The Project has undertaken various initiatives to provide best options for the sale of produce. The efforts to ensure best price for the produce are as follows.

i) Establishing Marketing Linkage

Timely information on marketing places, prices prevailing, quality requirements and associated aspects helped farmers to realize potential profit.

ii) Direct Sale by Producers

Many farmers / Farm women / Farm youth were encouraged to undertake sale of produces by themselves particularly watermelon, Ragi malt, Dhal, Nursery seedlings, Vermi compost, worms / manure etc. The system of sale by producers themselves did not stand in the long run in view of the following facts.

Almost all are nuclear type of families. Costly and inadequate transport is the biggest problem the great majority of the farmers are facing. As individual find difficulty to bargain.

The start of Producers association becomes inevitable and hence the initiatives were taken to start various associations to find sustainable solutions to producers' problems.

VI. Start of Producers Associations

In order to ensure minimizing cost of production, maximizing profit margin, undertake marketing and create additional employment opportunities, various associations were started. The Associations are expected to address end to end issues relating to a intervention/group of related intervention

Associations Functioning

- ☆ Rural Biofuel Growers Association, Hadonahalli - KMF
- ☆ Flower Growers Association, Hadonahalli – KSDH
- ☆ Jack Growers Association, Tubagere – KSDH
- ☆ Federation of SHGs of Tubagere Hobli, WCWD
- ☆ Organic Farming Farmers Association, Karnala – KSDA and APOF
- ☆ Chawki Rearing Centre, Heggadihalli – KSDS and CSRTI, Mysore
- ☆ Chawki Rearing Centre - Gangasandra, CSRTI, Mysore and KSDS
- ☆ Corn Growers Association, Hadonahalli

Associations in Pipeline

- ☆ Fruits and Vegetables Growers Association, Hadonahalli - KSDH
- ☆ Sericulture Growers Association, Hadonahalli – KSDS
- ☆ Fish Farmers Association

Outcome of the Project

In a span of three and a half years, the project is able to bring about significant impact in terms of shift in cropping pattern, net income, employment generation and improving social status. There was a shift in the cropping pattern from low value crops to high value crops, as well as Hi-tech interventions, introduction of subsidiary enterprises and value addition together enhanced opportunities for improving the income. The income during the benchmark period was Rs.3,96,74,431/- and it has increased to Rs.10,15,52,756/- during the project period. The net income generated was more than two folds during 2007-08, compared to bench mark period from among few selected interventions. The employment generated during the corresponding period due to those selected interventions was 2,52,875 man days per annum. The recognitions and awards for achievements and active participation in various commodity based associations have helped them to derive improved satisfaction in their vocations. The DBT Monitoring team, Government of India observed RBC-UAS (B) as one of the best models at all India level. The 5th NBDB meeting held on 14-07-2008 at New Delhi has advocated for replication of UAS (B), model for North Eastern States.

Case Study - A Successful Farm Entrepreneur

Sri Sadanand aged about 45 years educated up to 10th standard was working in a public sector company way back in 2000. He resigned for the job and took up farming in his two acre and twenty gunta land inherited from his ancestral property located at Tapasihalli, in Tubagere Hobli. The family consists of wife, son and two married daughters and he is residing on the farm. He was deriving a net income of Rs

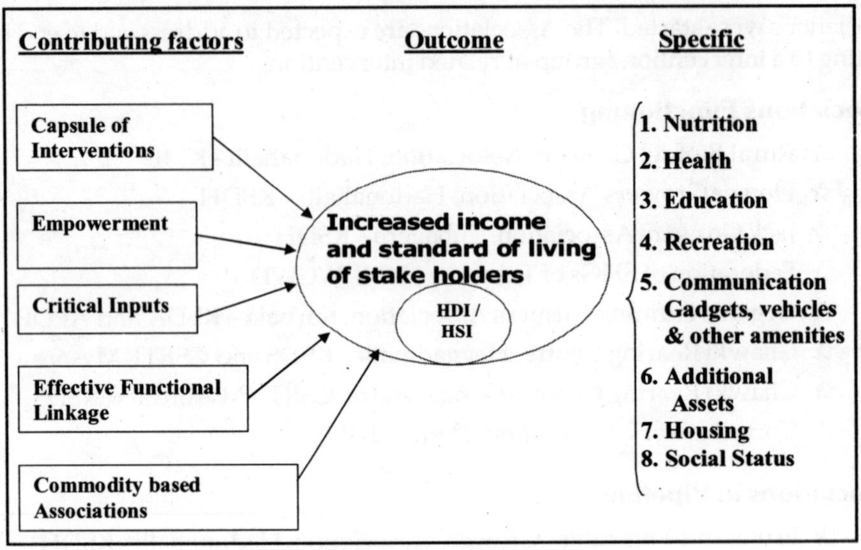

Figure 22.1: RBC Model UAS (B).

2.2 lakh during 2004-05. After Project staff intervention, the farmer is realizing a net income of Rs. 6.7 lakh per anuum as on 2008.

Conclusion

The farm families and rural people at large are gaining confidence and satisfaction in their vocations because of increased income and improved standard of living in the project area. Many visitors both with in and outside the country have visited project area, expressing great satisfaction on the achievements of the project.

Chapter 23

Innovations in Policy For Emerging Challenges of Agricultural Marketing for Enhancing Income

R.P. Singh

The supply pool for agricultural commodities is no more confined to domestic production alone. The countries having competitive edge in cost of production are looking for opportunities to dump their output anywhere in the world. Despite our complex and diversified agriculture, it is yet to pass the test for comparative advantage in terms of cost effectiveness for many important commodities. Unless, we succeed in inducing cost effectiveness having locational advantage, it may not be possible for us to harness the benefits of WTO through export of agricultural commodities. Besides, we have to face a number of emerging threats and challenges including unrealized opportunities in agricultural marketing; unwarranted imports by other countries: under exploited export opportunities for Indian products and distortion in domestic markets.

The term 'Agricultural Marketing Extension' (AME), often perceived as "Market Oriented Agricultural Extension" (MOAE), is quite different from "Marketing of Agricultural Extension" (MAE). While former implies broadening scope of existing agricultural extension system oriented towards enhancement of agricultural production in physical terms, the existing extension system has been devised and operationalized at beginning of planned era, with the sole objective of bridging the huge demand and supply gap existing with respect to various agricultural commodities. For all these years, the orientation remained more or less the same with changes in extension strategies in transferring technologies.

The last decade witnessed shift in development strategies leading to changes in consumer preferences and making alternative choices available to consumers. The consumer goods sector was quick to respond to these changes and introduced required re-orientation in strategies at macro level, but agricultural sector was very slow to respond in this respect which may be attributed to such factors as agro climatic conditions, soil-crop relationship, gestation period needed in evolution and dissemination of technological interventions and a host of other socio-economic factors.

Since India has entered in global market and signed WTO, in 1995, it became necessary to change our outiook on agricultural marketing system, especially in view of the Exim-policy the existing dynamics in domestic markets and the new agricultural policy. Market-oriented agricultural quality product has been the need of the day, which remained the real challenge for our farmers and public sector, dominated agricultural development system.

Growth - Policy and Agricultural Marketing

The current agricultural marketing policy framework and institutional structure is the outcome of different experimentations carried out at different points of time. The history of government intervention in agricultural marketing goes back to British time. The extent and form of interventions, however, changed over the period in response to problems experienced and solutions attempted.

Regulation of the practices of buying and selling was the first attempt introduced by the government for protecting interests of some sections of the society *viz.*, users of agro-raw materials, farmers and consumers. The overriding concern of agricultural marketing during pre-independence period to protect interests of the industry was substituted by concern for consumers as well as for farmers in post-independence era.

Though agricultural marketing is a State subject, the Government of India has been playing a crucial role in terms of laying down general policy framework, prescribing quality standards and providing technical and financial support to states for creating infrastructure and administrative set-up. Encouraging expansion of regulation programme, carried out commodity surveys, promoted grading and standardization of farm products and their derivatives, improved level of market information and intelligence network and upgraded the skills of marketing personnel by conducting training programmes for them. With the expansion in regulation programme, agricultural marketing boards were established and separate agricultural marketing departments were created in several states. Most of the wholesale assembling markets have been brought under the ambit of regulation programme.

The basic objective of these interventions in agricultural marketing was to ensure remunerative prices to growers and to prevent exploitation and malpractices by the intermediaries. These programmes paid rich dividend to agricultural sector in terms of hike in agricultural Production. Mechanism for sound agricultural marketing.

(i) Cost Effectiveness: The countiy's agriculture has to become more cost effective to meet the growing challenges and opportunities arising out of WTO agreements and the consequent of globalization impact. For this, future growth of agricuhure has not only be yield based but should be tilted towards ensured demand in the National and Internauonal markets.

(ii) Farmer- Agro- Industry- Consumer Linkages: There is a need to have strong linkage between agricultural marketing and agricultural production system as 'what to produce', how much to produce', 'how to produce' and 'for whom to produce.

(iii) Shift from Physical Output to Value Realization: Ensuing shift in the approach from increased physical output to increased value realization by the producers is the need of the time.

(iv) Agro- Processing and Value Addition of Agricultural Produce: India has great potential for expanding its export of processed products in view of the increasing production of fruits and vegetables, dairy products, meat and marine products, etc. The practice of contract farming is coming up around such processing centers. More and more processing facilities for such commodities need to be augmented.

(v) Grading and Packaging: In view of tremendous potential for export of fruits and vegetables, it is necessary to give greater attention to grading and packaging of these commodities by creating required infrastructural support.

(vi) Diversification of Agriculture: India has enormous inherent potentials for diversification and scope to introduce new areas in agricultural production such as hi- tech horticulture, precision farming, protected and organic farming, etc.

(vii) Augmented Infrastructural Facilities: A strong infrastructural support has to be developed in terms of cold storage chains, transport, credit support. market information and insurance to fully exploit opportunities for export.

(viii) Global Market Research and Information Center: An apex center at the national level for market research, international price analysis, global demand, availability and also to pass on this information to the appropriate authorities is very much required.

(ix) Marketing of Critical Agro-Inputs: For the marketing of agro-inputs like seed, it is necessary to streamline and simplify seed certification procedures, modify the Seed Act and enforce it strictly which should economically suit and benifit the farmers who is the first but unreached Besides, the other inputs such as fertilizers including organically produced fertilizers, pesticides, insecticides etc. required also such attention with greater participation of private sector, co-operatives, association and NGOs in the production and distribution of these inputs.

(x) Marketing Information Network: Easy access to timely and relevant information to farmers, traders, policy planners and other marketing agencies including stakeholders so as to enable them to have proper marketing decisions which is very much required. Even today appropriate application of IT in this area is still a misfit link.

(xi) Involvement of Private Sector: Efforts should be made to involve private sector in creating necessary marketing infrastructures under the atmosphere of liberalization and economic reforms in the contury, efforts should be made for strong public and private partnership.

Newly Emerging Areas in Agriculture for Export

There are several global issues and global dimensions, which the Indian agricultural sector has to confront as a consequence of globalization and liberalized trade regime. It is a reality that our gains through export of agricultural commodities depend on the promotion of those commodities for which we have a comparative advantage in terms of cost of production and quality aspects. Some of the new areas for the extension system to mould itself to perform in an effective manner are listed below:

(i) Green Food Production: It refers to organically grown crops, which are not exposed to any chemicals right from the stage of seed treatment to the final post harvest handling and processing.

(ii) Hi-Tech Horticulture: It is the deployment of any technology, which is modern, less environment dependent, capital intensive and has the capacity to improve the productivity and quality of horticultural crops.

(iii) Precision Farming: It implies to the application of technologies and principles to manage special and temporal variability associated with all the aspects of production and improving crop performance and environmental quality.

(iv) Sustainable Agriculture: It is the aspect of agriculture, which encompasses the use of natural resources- land, water and genetic endowment in such a manner that it should be technically feasible, economically viable, environmentally non-degrading and socially acceptable.

(v) Farming System Approach: The farming system refers to the activity mix of food crops, oil seeds, fodder crops, horticulture crops, livestock and other high value enterprises that would increase farm income while ensuring the goals of stability, sustainability/diversification and commercialization of agriculture.

(vi) Protected Cultivation: It is the method of harvesting the natural resources through the creation of an environment to enhance crop productivity by checking the vagaries of nature as weather variation often tend to have adverse effect on the yield and production of the crop.

(vii) Cold Storage/Cool Chain: It refers to the application of refrigeration technique in the preservation and storage of perishable agricultural products like fruits, vegetables and flowers to facilitate marketing system and to ensure higher return.

(viii) Micro-Propogation: It is a proven means of producing millions of identical plants under controlled and uniform condition, independent of seasonal constraints. It ensures economy in time/space and also gives greater output and higher income.

There are numerous other newly emerging concepts and techniques of crop production. The extension system will have to resolve these problems through its network in the years to come.

Agriculture Exim Policy

The Commerce Minister presented the new export-import policy for five years 1997-2002 on 31 March 1997. The main objectives of the policy were as follows.

- ☆ To accelerate the country's transition to a globally oriented vibrant economy to derive maximum benefits from expanding global market opportunities.
- ☆ To stimulate sustained economic growth by providing access to essential raw materials, intermediate components, consumables and capital goods required in augmenting production.
- ☆ To enhance technological strength and efficiency of Indian agriculture, industry and services and improving their competitiveness while generating new employment opportunities and encourage the attainment of internationally accepted standards of quality.
- ☆ To provide consumers with good quality products at reasonable prices.

EXIM policy 2002-07 contains policies in foreign trade *i.e.* with respect to import and export. Adequate attention has been given to export production. The Exim policy 2002-07 came into existence w.e.f. 1.4.02.

Agricultural Export Zones (AEZ)

It is a part of our EXIM policy. With a view to promote agricultural export from the country and remunerative returns to farming community in a sustained manner, AEZ was set up for end to end development for export of specific products from a geographically contiguous area. AEZ would be identified by the State Governments who may evolve a comprehensive package of services provided by all State Government agencies, State Agricultural Universities and all institutions and agencies of the Union Government for intensive delivery in these zones. Such services which would be managed and coordinated by State Government having provision of pre and post harvest treatment and operations, plant protection, processing, packaging, storage and related research and development etc. APEDA has been supplementing, through its schemes and provision, to State Governments and facilitating such exports.

Units in AEZ would be entitled for all facilities available for export of goods in terms of provisions of respective schemes. The Central Government encouraging manufacturers and exporters to attain internationally accepted standards and quality for their products.

WTO and Agricultural Marketing

There have been radical changes over years in marketing of agricultural commodities. So long as supply of agricultural commodities through domestic production did not outrun the demand, and free import was restricted, producers enjoyed a fair degree of advantage in freely selling their products at remunerative prices amidst a strong price protection mechanism through minimum support price policies. Occasional problems in marketing of agricultural commodities like sugarcane, vegetables and coarse cereals were reported in different states but these were resolved with mass movement and timely political intervention. Thus, the production- oriented agricultural marketing dominated the country for quite a long period. However, the question pertaining to realization of remunerative prices and net gains from agriculture continued to be disputable for all these years.

There is a great need to reorient our agricultural production systems and import-export policy for enhancing export promotion by considering followings:

(i) Promoting of eco-friendly inputs, precision agriculture practices and participatory technology development for sustainable increase in productivity and production. In order to increase the exportable surplus.

(ii) Exploit the agro-ecological diversity of the country, for production and export of agricultural produce from different regions during the off-seasons in importing countries. The APEDA and SAD's can work on this and the State Departments of Agriculture, Horticulture, Agricultural Marketing and State Agricultural Marketing Boards can implement this.

(iii) Enhance the competitiveness of the Indian agricultural produces; cost of production must be brought down by inputs management and increasing yield per unit of inputs through facilitating adoption of latest practices and reducing the yield gaps.

(iv) Training must be given to farmers on hi-tech modules, pre-harvest management and post-harvest handling of produce to match importing country standards.

(v) Agriculture must be treated as agri-business and all incentives must be extended for its promotion and value added products from agricultural commodities in general and dry land crops in particular.

(vi) Credit availability must be enhanced and all procedural bottlenecks must be removed the institutional lending, besides, creating awareness among farmers to use crop insurance programmes available.

(vii) Harness benefits of new scientific advances such as bio-technology, cloning, remote sensing, modeling, information and IT for farmers and rural communities through Internet aided market oriented extension networks have to be materialized that too in local language.

(viii) Farm Science Centers (KVKs) should be strengthened as nerve centre for dissemination of technical know-how and market intelligence. Besides, promoting the market export oriented quality produce of the commodity.

(ix) Formation of commodity specific extension kiosk on indigenous food, horticultural crops, livestock, poultry, bee keeping, mushroom, medicinal and aromatic plants, sericulture, tea and other value addition enterprises.

(x) Distance education for farmers and small agri-entrepreneurs need to be explored and strengthened through television, interactive audio and video system, besides. print and learning materials which would be the distinguishing feature of learning process of agricultural marketing system.

(xi) Launching awareness training programmes on import-export policy for producers, sellers, consumers and stakeholder.

(xii) The knowledge on freely exportable and importable commodities, restricted as well as non-exportable/Importable at various levels as it has implications on the marketing of agricultural products, which needs application of Exim-policy.

Changing Context of Agricultural Extension

Challenges offered by sustainability (depleting natural resources), and trade fronts, World Trade Organization (WTO) agreement, the changing nature of agricultural technology (from public to private goods), rapid developments in IT and a changing development agenda (stakeholder participation, decentralization and faster reduction of poverty) have prompted a re-evaluation of the role of extension in many countries. Farmers need to be supported with information, knowledge and skills to adopt improved technologies that result in improved farming characterized as productivity enhancing, vulnerability reducing, employment creating and surplus generating for improving the livelihood in a sustainable fashion. However, the requirements of farmers and rural families go beyond agricultural production technologies following increasing penetration of markets into rural areas and the emergent need to tailor products to stringent market requirements.

Role of HRD in Marketing

There is a need to train the farmers in the areas like product planning, marketing information. Preparation of quality produce for marketing, improved marketing practices, rules and regulations, input marketing, etc. We have strong network of Extension in the form of ICAR and SAUs; Extension system of ministry of Agriculture and Rural Development, extension work done by the Non- Government Organizations (NGOs); Business Houses and voluntary agencies. The rapid advancements in the technological front necessitate that these should reach to the communicators in time so that they in turn could disseminate it among the producer. The transfer of technology being a complex and continuous process, there is no alternative but to develop human resources to achieve desired goal. The first line as well as development extension agencies, both at center and in the states, involved in Extension Education activities related to market and import-export policies, required to pay added attention towards market oriented transfer of technology for reducing the operational cost and increasing the margin of profit.

Conclusion

The agricultural extension strategy initiated to achieve self-reliance in agriculture has to be reoriented and made market centered. Such a shift in the strategy has become inevitable owing mainly to strong backward linkage that agricultural marketing has with agricultural production leading to - as 'what to produce' and how much to produce, ensuing shift in the approach in accordance to declared EXIM policy (2002-07) by the Government of India listing freely exportable and importable commodities (restricted as well as non-exportable/importable commodities) and lack of accessibility to Information Technologies due to illiteracy, knowledge ect.

It is well recognized that research and extension scientists have complementary roles in agricultural development. The success of each group is determined largely by the effectiveness of linked activities, therefore, efforts should be directed to improve the existing linkage between farmers- research/extension. Transfer of appropriate knowledge pertaining to meet the emerging challenges of marketing for enhancing agricultural income will be more effective if all of them show more commitment on their respective fields and co-ordination among them. Involvement of the farmers at planning level may accomplish the execution of the various activities.

In a competitive environment of today and in the wake of missing expectations of the consumers. the hall-mark of coming decades will quality, innovations and efficiency in every sphere of business activity. By setting standards and enforcing commercial laws, government has to foster economic growth and provide the framework within which a modern economy can function. It is high time that we take stock of the existing situation, improve upon it, adopt and adapt it according to the exigencies of the circumstances. We are running a race and we must run faster in order to improve our relative position at the right path for meeting the challenges of global competitiveness in the agricultural marketing sector for enhancing farmer's income who is the first.

Chapter 24

Changing Agri-Rural Environment: Potential and Priorities

A.K. Singh, Jitendra Chauhan and R. Roy Burman

Agriculture provides sustenance to two-third of India's population, which has crossed more than one billion and still growing. With decreasing share of agriculture to GDP is resulting into rural poverty and people are faced with the livelihood problem. The rural poverty and livelihood issue is aggravated by rising standards and cost of living at one end and on the other shrinking land holdings, steadily rising input costs, globalization and associated liberalization of markets and demand of competitiveness in unit cost of production, quality and ability to market produce and primary processed products without excessive post-harvest losses. Factors of demand and supply, poor PH-infrastructure, markets and the forces that intervene the agriculture markets usually create slump in the market price whenever there is a bumper harvest of any crop or commodity. Orchestrated distress sales, denying growers remunerative prices, compels the farmers to shift over to the crop and commodity, which in turn creates scarcity and soaring consumer prices. A vicious cycle that does not seem to be ever ending to the detriment of growers and consumers' interest.

Poverty in India is essentially a rural phenomenon particularly among the rural poor. Nearly 70 per cent of the poor live in rural India and bulk of them are farmers and agriculture dependent. An imbalance exists among total production, national food security and household besides individual level food security. The impact of climate change on agriculture could result in problems with food security and may threaten the livelihood activities upon which much of the population depends. Climate change can affect crop yields (both positively and negatively), as well as the types of crops that can be grown in certain areas, by impacting agricultural inputs such as water for irrigation, amounts of solar radiation that affect plant growth, as well as the prevalence of pests. Indian climate is dominated by the south-west monsoon, which brings most of the precipitation. It is critical for the availability of drinking water and irrigation for agriculture. Agricultural productivity is sensitive to two broad classes of climate-induced effects; (1) direct effects from changes in temperature, precipitation, or carbon dioxide concentrations, and (2) indirect effects through changes in soil moisture and the distribution and frequency of infestation by pests and diseases. Rice and wheat yields could decline considerably with climatic changes. However, the vulnerability of agricultural production to climate change depends not only on the physiological response of the affected plant, but also on the ability of the affected socio-economic systems of production to cope with changes in yield, as well as with changes in the frequency of droughts or floods. The adaptability of farmers in India is severely restricted by the heavy reliance on natural factors and the lack of complementary inputs and institutional support systems. Extensive and dramatic changes are occurring in the Indian economy as a result of the globalization of economic activity and the implementation of structural adjustment reforms. Indian agriculture faces the dual challenge of feeding a billion people in a changing climatic and economic scenario. However, there is not much scope to increase the area under food grains. Switching to other crops is also difficult in regions like West Bengal, where three rice crops are grown in succession through the year. Agriculture is the predominant means of livelihood for a large number of peasant cultivators and agricultural labourers, for whom it is not easy to shift to other occupations. Due to their low financial and technological adaptability, such groups are potentially vulnerable to both climatic changes as well as economic change.

Hence, it becomes important to examine the vulnerability to climate change not in isolation but in the context of ongoing economic changes. This translates into a dynamic interpretation of vulnerability and has important implications for developing adaptation strategies.

The agriculture will still continue to be the engine of the national growth and development. Rapid and balanced growth of agriculture is thus essential not only to achieve self-reliance at national level but also for household food security and to bring about equity in distribution of income and wealth resulting in rapid reduction in poverty among the rural poor. The Green Revolution process based on the synergy of advanced technology, effective policies and desired political will ushered in the mid 1960's, followed by white, yellow and blue revolutions, almost tripled the agricultural production, halved the real price of major food grains and other food items, increased the per capita food consumption by 30 per cent, more than doubled per capita GDP, and also halved the percentages of undernourished and rural poor. India is one of the world leaders in the production of milk, rice, wheat, fruits, vegetables, cotton, sugarcane, poultry and fish. It is poised to achieve an all-time high of 220 million tons of food grain production.

India has the largest cattle and buffalo population, second largest population of goats and the third largest of sheep in the world. Total livestock population is around 495 million. We are also the largest milk producers (91 mt) in the world today. Yet, we have not been able to compete globally in the export of milk products, meat and even live animals- as being done by Australia, New Zealand, Holland etc. Today, we also produce 41 billion eggs and have 490 million poultry birds. We need to link farmers to markets – the way it is done in Gujrat – through exemplary cooperative movement. Such efforts need to be replicated fast in other states as well. Supply of fodder and feed, use of silvi- pastoral approach in rainfed areas (especially Rajasthan, Madhya Pradesh, Maharashtra, Andhra Pradesh etc.), establishment of A.I Centers, Livestock, Clinics, Supply of good quality, vaccines, insurance of livestock (at least of valuable productive ones), establishment of modern abattoirs, processing, packaging, storage, and marketing (including export) facilities would all lead to much faster growth in livestock sector- for which India has great comparative advantage but so far not fully exploited. For example, we should be major producer of Mozarella Cheese, being the largest buffalo milk producer, yet our share in global market is almost negligible.

Another important issue is regarding crop-livestock integration, especially in arid regions. In some areas, such as Rajasthan and Maharashtra, major thrust should be on silvi pastoral practices using agro-forestry and use of rangeland pastures and legumes, beside, drought tolerant shrubs and trees such as Khejri (*Prosopis*), babool (*Acacia*), etc. In Rajasthan Canal area, especially Bikaner and Jaiselmer Divisions, a re-look is necessary at arable cropping to be replaced by raising pastures and livestock considering the long-term sustainability and profitability. Appropriate mid course correction through needed policy reforms would be desirable. Also we need to protect and improve our local breeds- being most valuable assets.

Most critical factor for faster growth in future will be the input use efficiency. It would demand for timely operations/precision farming. Integrated Natural Resource Management (INRM) would call for more rational use of water, seeds, fertilizers, pesticides etc. Precision farming through greater emphasis on mechanization (both, in irrigated and rainfed farming systems) would be the major recourse in future. Large-scale manufacture of equipment and machinery/tools (such as zero till drill, planters, seed and fertilizer drills, sprinklers, mechanical harvesters, combines and small tools for weeding, harvesting and threshing/cleaning etc.) and their easy availability at appropriate cost will help in accelerating future growth in agriculture. Hence, this aspect being in the national interest, should receive priority attention during the XIth plan.

Overtime Changes in Extension Methodologies and Approaches

Agricultural extension has undergone considerable change over several decades. New policies are resulting in a large proportion of extension now being undertaken by the private sector also. These changes are continuing and have presented the agricultural community with certain challenges and problems. Traditionally, the major aim of agricultural extension was to advance the transfer of new techniques, new knowledge and new technology from the research sector to the farmers. The linear and mainly technology-driven model reflects the modernistic development perspective of the 1960s and is based on the positivist science paradigm. An example of the Transfer of Technology is the green revolution of the 1970s. The green revolution packages were suitable mainly to areas of high natural potential having uniform and controllable growing conditions. This model, aiming at a widespread adoption of technologies, is likely to be successful in relatively homogenous, low-risk, natural and social environments, where farmers live under similar conditions, perceive the same kinds of challenges and share a common set of beliefs and values. In Farming Systems led more emphasis is laid on (contractual and consultative) farmers' participation for better understand their complex situation and the inter-dependencies among elements of farming systems in order to develop adapted technologies (Biggs, 1989; Farrington and Martin, 1987; Rhoades and Booth, 1982).

Analytical the first step is to identify the main issues in the changing landscape of agricultural innovation and development that we need to tackle. What are the main changes that affect agricultural R&D, extension, education and agricultural development in general? Once we have identified these changes, can we ask further questions about how each component and actor in the system has to change? These questions may have to do with the nature and extent of change, possible period or sequencing of change, or partners in change, new norms or rules (such as prioritization, accountability, incentive and reward structure, natural resource accounting and poverty impacts).

The agricultural extension system in our country was devised to support the multifaceted approach for achieving the goals of self-reliance in the sector. The agencies, directly or indirectly associated with the task, include national institutes/ directorates, SAU's network, input companies, NGOs, agro-processor, private consultants, farmers associations and other organizations. The transfer of technology, human resource development and information and communication system formed the key components of extension strategies. A re-evaluation of the role of extension and restating the strategic approaches to agricultural extension was necessitated due to the challenges offered by sustainability considerations, complex problems arising form the Agreement on Agriculture under WTO, the changing nature of agricultural technology, rapid developments in Information and Communication Technology (ICT) and the changing development agenda. Public expenditure on agricultural extension and its control have been justified on the ground that support for agriculture leads to reduced food prices and increased food security, poverty alleviation, employment generation, environmental conservation, etc. which benefit the whole population specially the poorest of the poor in the society. Important and pertinent questions are raised in this connection as to who should fund and deliver extension particularly agricultural extension which has to play various roles ranging from provider to coordinator, facilitator, arbitrator, regulator, guarantor, etc.

Farmers need to be supported with information, knowledge and the skills to adopt improved technologies that would enhance productivity, employment opportunities and sustainability. Therefore, the ambit of agricultural extension has remarkably widened and the onus of managing the complex affairs rests with agricultural scientists and extension specialists working with public and private sectors as well as NGO's of the country so as to help the farmers to produce according to the requirements of the markets.

The supply pool for agricultural commodities is no more confined to domestic production alone. The countries having competitive edge in cost of production are looking for opportunities to dump their output anywhere in the world. Despite our complex and diversified agriculture, it is yet to pass the test for comparative advantage in terms of cost effectiveness for many important commodities. Unless, we succeed in inducing cost effectiveness having locational advantage, it may not be possible for us to harness the benefits of WTO through export of agricultural commodities. Besides, we have to face a number of emerging threats including unrealized opportunities in agricultural marketing; unwarranted imports by other countries; under exploited export opportunities for Indian products and distortion in domestic markets.

Since India has entered in global market and signed WTO, it is necessary to change our outlook on agricultural marketing system, especially in view of the Exim-policy and the existing dynamics in domestic markets and the new agricultural policy. Market- oriented agricultural extension is the need of the time, which is the real challenge for our public sector dominated agricultural extension system. Following factors are to be taken into consideration for a sound agricultural marketing extension for the sustainable agricultural development:

- ✰ Farmer- Agro- Industry- Consumer Linkages: There is a need to have strong linkage between agricultural marketing and agricultural production system as 'what to produce', how much to produce', 'how to produce' and 'for whom to produce'.
- ✰ Shift in Physical Output Value Realization: Ensuing shift in the approach from increased physical output to increased value realization by the producers in the need of the time.
- ✰ Agro- Processing and Value Addition of Agricultural Produce: India has great potential for expanding its export of processed products in view of the increasing production of fruits and vegetables, dairy products, meat and marine products, etc. The practice of contract farming is coming up around such processing centers. More and more processing facilities for such commodities need to be augmented.
- ✰ Grading and Packaging: In view of tremendous potential for export of fruits and vegetables, it is necessary to give greater attention to grading and packaging of these commodities by creating required infrastructural support.
- ✰ Diversification of Agriculture: India has enormous inherent potentials for diversification and scope to introduce new areas in agricultural production such as hi-tech horticulture, precision farming, organic farming, etc.
- ✰ Augmented Infrastructural Facilities: A strong infrastructural support has to be developed in terms of cold storage chains, transport, credit support, market information and insurance to fully exploit opportunities for export.
- ✰ Cost Effectiveness: The country's agriculture has to become more cost effective to meet the growing challenges and opportunities arising out of WTO agreements and the consequent globalization impacts. For this, future growth of agriculture has not only be yield based but should be titled towards ensured demand in the national and international market.

- ☆ Export Promotion: There is a great need to orient our agricultural production system for enhanced export promotion.
- ☆ Global Market Research and Information: An apex center at the national level for market research, international price analysis, global demand, availability and also to pass on this information to the concerning authorities is very much required.
- ☆ Upgrading of Human Resource through Trainings: There is a need to train the farmers in the areas like product planning, marketing information, preparation of produce for marketing, improved marketing practices, rules and regulations, input marketing, etc.
- ☆ Involvement of Private Sector: Efforts should be made to involve private sector in creating necessary marketing infrastructures.
- ☆ Marketing Information Network: Easy access to timely and relevant information to farmers, traders, policy planners and other marketing agencies so as to enable them to have proper marketing decisions is very much needed. Application of IT in this area is still a miss-link.
- ☆ Marketing of Agro- Inputs: For the marketing agro-inputs like seeds, it is necessary to streamline and simplify seed certification procedures, modify the Seed Act and enforce it strictly. Greater participation of private sector, coo-operatives and NGOs in the production and distribution of seeds is required.
- ☆ Declared EXIM Policy (2002-07) by Government of India: The knowledge on freely exportable and importable commodities, restricted as well as non-exportable/importable at various levels as it has implications on the marketing of agricultural products.

Effect of Rural Migration and Urbanization on Agriculture

Rural population has migrated to cities due to lack of employment in agriculture sector as livelihood security of rural poor is uncertain in this sector, hence there is need to empower rural women to secure their livelihood through other alternatives. Livelihood of rural poor can be secured by way of adopting agro based highly remunerative and income generating enterprise like Mushroom production, IPM, Processing and preservation, Vermi composting, Nutritional garden, Dairying, Poultry, Strawberry cultivation and value addition etc. The intervention like crop enterprise technology, approach and service diversification can contribute significantly in sustaining livelihood.

In advance countries like USA there has not only been a substantial increase in employment of women but the last decade has seen the number of self employed women increased by 9 per cent as compared to 13 per cent in the number of self employed men (D. Lalitha Rani). This trend depicts that entrepreneurs seem to be ideal for women who would like to participate in the country's socio- economic development. Moreover, the role and degree of integration of women in economic development is an indicator of women's socio- economic dependence and social status, therefore, immense need is felt by developing nations to mobilize women into different professional fields including entrepreneurship in a planned manner.

The economic development of a country is correlated with the population, employment, literacy and effective utilization of resources together with industrialization process, which accelerates the growth including social change. The mobilization of economic resources of the country is highly beneficial by the technological advancement depending on human resources development, which has gained momentum in the recent past. Human resources comprise both men and women.

For all round development and empowerment of women and for reducing gender inequality, it is imperative to disseminate right kind of relevant information to the society in general and women in particular. Empowerment of women can become a reality only when they have the confidence and ability to engage in independent and income generating activities and enterprises. The country needs to mobilize and utilize fully, all its resources, including human resources. The participation of women in economic activities is necessary not only from a human resource point of view but is essential even for the objective of raising the status of women in the society. Although, women form almost half of the world's population and form an integral part of the society, yet, women world over have been suffering from various types of discrimination and deprivation from times immemorial, owing to socio-cultural ostracism and other ethos. During the last few decades, there has been a global concern to bring women interested in mainstream of economic activities so as to ensure their quality and empowerment.

Food processing helps in promoting a synergy between the two major components of the economy, namely, the agriculture and industrial sectors. Strengthening the agri-industrial base of the country will also be beneficial in preventing population shift from rural areas to urban areas, improving economic condition of the rural people, promoting entrepreneurship and in generating employment in rural areas. Having become aware of the vast potential of promoting agri-business in rural India. Some of the corporate houses have taken a lead by establishing processing units and strengthening marketing activities in rural areas and the initial response seems to be a quite encouraging.

Growers have no share in value-addition to their produce and by-products. Large part of year of their family labour is un or under employed. Landless often migrate in search of wage earning dislocating education of children. It has brought post-harvest technology and agro-processing at the centre stage. Scientists, policy makers and development agencies are looking towards rural food and agro-processing as a tool for additional income and employment to the rural people in rural areas having potential scope for enabling them to meet their needs at the least cost, retain and use waste and by-products usually to feed, fuel and industrial raw material value in the production catchments and market surpluses after value addition. Leading luminaries of Indian agriculture consider PHT and Value Addition as half the agriculture story.

In the developed countries of the West, the rural to urban migration and the urbanization are associated with a vertical shift in the labour force from the agricultural sector to the urbanised-industrial sector, whereas in the developing regions as India, migration is from rural agricultural sectors to urban informal sectors. This is a typical manifestation of under-development, poverty, and spatial disorganisation of economy of underdeveloped sectors of the society, which arose partially as a result of past colonisation and its adverse consequences on space economy. These displacements are not due to structural changes within the labour force as seen in the West, but as a dislocation of uprooted workers and peasants from the marginalised countryside to involuted urban centres. This displacement in India is a typical characteristic of urban growth that has outpaced industrialisation. It is a system of underdevelopment, and it tends to compound further underdevelopment (Mukherji S. 2002)

There is overwhelming evidence that internal migration can lead to positive change in both sending and receiving areas (Deshingkar and Grimm, 2004). Migration can help to reduce poverty or to halt the slide into poverty. For example, the income earned by women migrating from West Bengal to Delhi to work as housemaids meant that they had been able to move out of poverty (Mukherjee 2004). It can increase income, savings and assets. For example, a study of 955 migrant households in Tamil Nadu found that 57 per cent of lower income migrants had seen their income increase and 53 per cent had increased their asset holdings (Sundari, 2005) and migrants from Mahbubnagar District, Andhra Pradesh to the paddy fields of Karnataka have been found to save an average Rs 2000-3000 per season (Khandelwal, 2002). It can lead to the sending of remittances to marginalised sending localities, which can be used to invest in human capital or in productive assets in sending localities and can play an important role in reducing vulnerability, improving food security, stimulating land markets in sending areas, increasing local wages and the demand for local goods and services and generally improving the economy (Deshingkar, 2006). It also helps tighten rural labour markets (Wiggins and Deshingkar, 2007).

There are some negative impacts to acknowledge, however, migration can lead to an acute shortage of labour and high dependency ratios in sending areas. Mass male migration can lead to worsening poverty, but these risks are off-set where wage rates are sufficiently high to allow regular remittances. People who are away for a long time may lose access to natural resources and lose their voice in community decision-making. Migration can also have a negative effect on collective action and natural resource management, where significant labour inputs are required (Deshingkar, 2003).

Emerging Innovative Methodologies/Models/Cases of Extension Education

Today, the transfer of technology model is often viewed as the antithesis of participatory research. However, this is often not the case. In fact, much of the present participatory practices can still be classified as an expansion of the transfer of technology model because information is obtained from farmers and incorporated into scientific research. Participatory methods are used to better meet farmers' needs and to adapt technologies to site-specific circumstances at a relatively late stage of the research process. By the mid-1980s, people were re-thinking the transfer of technology model and emphasis was on the farmer. There are different types of approaches summarized under 'Farmers First'. Farmers became part of the process of generating, testing and evaluating technologies that promoted sustainable agricultural production. The main outcome expected from these approaches is the generation and adoption of new, appropriate technologies by small, resource-poor farmers to aid in solving production constraints in order to increase farm productivity and income (Selener, 1997). Most of the current NRM research initiatives focus on the generation and provision of technologies, assume a functioning linear research-development continuum, use mostly consultative forms of participation, and consider participatory research as a tool for applied and adaptive research. Therefore, they principally fall into the categories of 'transfer of technology' and 'farmers first' approaches. Longer-term participatory learning and action research approaches are only beginning to be chosen by international agricultural research centres (IARCs) as they require a different kind of professionalism and challenge the mandate, *i.e.*, they are considered to fall under the sphere of development rather than research.

The present departmental mode of organization and management public systems in agricultural research and extension need to be moulded in an innovation system framework. This implies demolishing dichotomies like research/extension, plan/non-plan, centre/state, public/private, agricultural/rural, and so on. More effective mechanisms must be identified and developed to internalize stakeholder involvement in decision-making and improve partnerships across the board. There is need to shift from an hierarchical and linear technology generation-diffusion model to a non-linear and holistic learning mode.

This will necessitate creating and nurturing a broad-base 'think-tank' at the ICAR and SAU levels. The model of the Technical Advisory Committee (TAC) of the CGIAR with its own secretariat and loop in the funding process may be adopted with suitable modifications. At the level of institutes also, there is need to expand and back up bodies like the Research Advisory Committee (RAC), governing Councils etc. to play this role effectively.

Bureaucratic and financial norms of functioning and accountability processes have proven inadequate for meeting scientific goals. These need to be replaced by more autonomous, flexible and performance-centered processes and management climate.

- ☆ Improving Extension Research Linkages: The reforms in agricultural extension on the aspects of research extension linkages include promotion of direct interface between farmers and scientists to minimize transmission losses. Activating existing interface mechanism and institutional linkages through various forum and research prioritization based on strategic research and extension plans are the means suggested for improving research extension linkages.
- ☆ Capacity Building of Extension Functionaries: Various measures are suggested for capacity building of extension functionaries include

formulation of HRD policy by states to avail central support to HRD and formulation of a long term training plan for extension functionaries. Such training plan covers two major aspects, namely foundation courses and professional courses. The foundation courses may cover areas like need assessment technique, group formation, development of entrepreneurial skills Agri-business, Agri-business management, WTO and its implication, marketing of agricultural production, post- harvest management, management of common property resources, use of different type of media, communication, project preparation, data collection, analysis and documentation, etc. The other steps suggested for capacity building are (i) one time catch up grant for developing infrastructure for training (ii) upgrading of state level extension management training institutions (iii) developing professionalism in cost-effective manner (iv) strengthening the role of MANAGE and (v) networking among all SAUs.

- ☆ Empowerment of Farmers: In order to project farmers as major stakeholders of the extension system, it was suggested to involve farmers in setting extension system, it was suggested to involve farmers in setting extension agenda and also in the implementation of programmes through farmers user group. Contracting arrangements between government agencies, organizations for extension services and farmers have also been suggested. Acquisition of skill by farmers through training and other programmes also form integral component for empowerment of farmers.
- ☆ Main Streaming of Women in Agriculture: The necessity of targeting women to receive information related to their work is highlighted. Special programmes for improving access to extension and training by rural women farmers are the other measures suggested in this direction.
- ☆ Use of Information Technology: Increased use of information technology in agricultural extension has been advocated. It is pinpointed that information technology can be effectively applied in agricultural marketing extension. Farmers participation in IT programme, support to develop information technology for agricultural extension by various states, development of private information shop, effective use of IT, etc. are the other important aspects covered under the reforms in information technology. With the launching of 'Kisan T.V Channel' and Call Center Services the relevance of IT in farm information has become more pronounced.
- ☆ Financial Sustainability and Resource Mobilization: Realizing that public funded extension will continue to play a predominant role in technology dissemination, cost effective mechanisms for extension services has been suggested. The steps suggested for financial sustainability include (i) efficient use of available resources (ii) privatization of agro- services (iii) realistic cost of recovery of agro- services (iv) co- financing of public extension by farmers and (v) initiating new financial system.
- ☆ Changing Role of Government: The states will have to play an effective role for the regulation and enforcement of the extension system. Enhanced competition among extension agencies, strengthening of farmers association

and development of physical infrastructure are the other parameters to be taken care of by the state government.

The extension system with the involvement of Government agencies, NGOs, farmers organizations, private sector agencies, para workers, etc is going to be more effective and dynamic. The agricultural extension system will have to transform itself with the capacity building to meet the challenges arising out of farming system approach. The transfer of technology system is likely to undergo radical reforms as the farmers need a wide range of services on aspects like marketing, credit insurance, infrastructure (including cold- chains), entrepreneurship, etc. It is possible only through changes in the institutional and organizational set up of our extension system. The extension system has to be sent on a "system management" mode, since the goal of extension has to make a shift from "technology dissemination" to "system management".

As high competitiveness is a priority in the globalized world and much of it, is underpinned by information revolution, increasingly knowledge will substitute monetary inputs. Technology packages largely will become knowledge-based. The existing agricultural technology development and transfer system in India, including the human resources and infrastructures, are outmoded, causing serious technology gaps and slippage. The extension and technology transfer system must be updated and the huge urban-rural digital divide must be narrowed. A crash training and skill development programme and establishment programme and establishment of rural information centers must assume high priority. Such programmes would particularly be attractive to the rural youth, who are mostly unemployed or underemployed.

Introduction of para-professional/vocational courses in specific, but applied areas of agriculture for grass root level services to limit the undesirable intake in higher education and also to support self-employed graduates in areas like Agri-business and Agri clinics must from part of our strategy for employment oriented agricultural education. Such para professionals should have the advantage of fluency and flare to communicate in local language; base the line information on village life, livelihood means, soils, water, flora and fauna (domestic and wild), traditional practices and beliefs and modern knowledge and skills on diverse crops, livestock including their varieties and breeds, agronomic practices, methods of soil and water conservation and testing; necessary information on ecological principles in managing diverse natural resources, practical exposure to general field problems and experience to administer prescribed treatments.

It is unaffordable to refrain ourselves from reorienting agricultural education in consonance with contemporary and futuristic needs of job markets. The foremost task is to regulate number of admissions in subjects and disciplines based upon regular reviews and assessment of job markets and corresponding manpower needs. In order to foresee the future employment avenues it is necessary reengineer agricultural education in consonance with requirements of various stakeholders *i.e.*, public institutions, private enterprises, graduates seeking self-employment, NGO's and farming community. Keeping the focus on job market in view, the introduction of subjects and methods of teaching and learning aligned fully with employability should from the core of the strategy on new courses of agricultural education. While making the blue print to reengineer the course curricula, there is a great need to orient it with other sectors of the economy, country's national commitments and international obligations. The course curricula must find adequate importance on newly emerging frontier technologies, global concerns like sustainability, food security, livelihood security, poverty alleviation, biodiversity and conservation, natural resource management, environmental safety and new areas like farming systems, hi-tech horticulture, precision farming, green cultivation, Agri-business, international trade and regulation and other enabling aspects.

The curricula must be framed in such a manner that the graduate professionals must have fluency to convincingly communicate in local language; expertise and upto date knowledge on production, protection, processing marketing and managerial aspects of agriculture and allied areas. Entrepreneurship ability, capacity to take initiatives in new ventures; business acumen, marketing skills and competence to infuse primary processing activities (grading and labeling) so that produce earns maximum price and profitability; leadership and team working, managerial qualities, computer literacy to update technical management and commercial knowledge and national policies and global happenings to offer latest and viable information on and solutions for problems faced by farming community must be the added focus of agricultural education. Subsequently, suitable revision of post-graduate courses in terms of subject matter knowledge as well as capability and willingness to work in inter and multi-disciplinary team and ability to build and lead research teams should follows.

With changing accent of education on producing professionals suiting market needs, the teaching methodology will require a radical change. Envisaged strategy is to move from teaching to learning with increasing emphasis on hands-on practice in real life situations infusing self-learning. A life-size and level ground for learning will require creation of facilities and enabling environment for practice. Strategy towards this goal may require to forge intensive tie-ups with development departments, industrial/business houses and NGOs enabling students using their facilities and expertise for practical training.

Relevance and excellence must underpin the functioning of our universities. None other than Nobel Laureate Professor Amartya Sen had stated on November 14, 1999 that "University education in India is in a state of crisis. It is not a crisis of lack of resources, it is deterioration of quality quite a lot of our success depend on not what we inherit but on what we acquire and how we apply it". What is needed today is extra-ordinary teachers and scientists who will be able to face challenges of future agriculture and produce the product which will be competitive in the international market. For this, universities must nurture and nourish extra-ordinary teachers and scientists devoted to the cause of science and at the same time are dedicated to their profession of teaching, research and extension.

In order to make way for teaching in real life subject and expanding room for practice and innovativeness through hands-on training it will also be necessary to build competence and capability of teaching faculty in emerging areas of science and technology and skills in vocational courses. The periodic refresher programmes for knowledge upgradation of teaching faculty must be ensured through centre of excellence in various professional areas. In fact, investment to develop faculty is a key strategy to facilitate change through introduction of new subject of learning, optimally utilize the institutional infrastructure for practical training and forge linkages with other sector of economy for skill development during internship.

Improving Environment for Different Stakeholders

There is a felt need for insights into the relationship between agricultural science and policy in the Indian context. While the latest National Agricultural Policy document of the Government of India provides an immediate focus, there are several questions concerning the capacity of R&D to identify and respond to the crucial and durable elements of agricultural policy on the one hand, and to contribute effectively to policy formulation it self on the other. The nature, extent and direction of changes in agricultural R&D, education, and extension demand detailed deliberations among the stakeholders – the community of professionals, farmers, NGOs, policy makers; both from private and public sectors. Other changes include increasing private sector presence and profit motives in agricultural R&D on erosion of public sector commitment to basic and poverty oriented research, and the potential and uncertainties of emerging biotechnology and information technology regimes. The demand for sustainable agriculture and poverty alleviation, the need for introspection and evaluation in R&D, are other challenges demanding critical changes in R&D paradigms.

Funding for public agricultural research extension and education continues to be stressed, more so at the state level, with conflicting policy signals. The government makes a commitment to raise public investment in R&D on the one hand; it enjoins ICAR and SAUs to mobilize their own resources on the other. This imparts uncertainly and inefficiency. It is high time that the issue of the level of investments, roles of public, private and international players, are discussed and stored out in a consultative mode. Past efforts have been partial, ad hoc and ineffective.

There is a need for enhanced social science research capabilities both for policy analysis and to develop R&D strategies for specific agro-ecological regions and farming systems. There is need for greater interaction between natural and social sciences. This will help the R&D system to build bridges with policy making on the one hand and the endures (farms) on the other hand in interactive ways. The National Agricultural Policy visualizes technological upgradation of Indian agriculture as a core element of the agricultural development strategy. The goals of sustainable agriculture, food and nutrition security, risk management as well as growth and agricultural trade depended critically on improved R&D processor in the sector. Specifically, it endorses the concept of regionalization of agricultural research based on identified agro-eco regions, (location-specifically). Use of frontier sciences, participatory and proprietary approaches in R&D, strengthening research-extension linkages, and a "well-organised, efficient and result-oriented agricultural research and education system to introduce technological change in Indian Agriculture". It calls for a redefinition of the partnership between central and state governments, assigning an upstream role for the centre. It also charts a pathway of incentives, support systems, investment imperatives and policy to "actualise that vast untapped potential of Indian agriculture".

There is increasing recognition that farmers obtain technical information from a wide range of sources, that such information has to be adapted to a variety of agro-ecological and socio-economic conditions, and that farmers' own knowledge is essential in the selection and adaptation processes. The boundary between public sector extension services and farmers is shifting: farmers are reaching higher into the technology generation and transfer system in order to 'draw down' suitable technologies. The boundary is also becoming more uneven: those producing commercial crops are obtaining technical information from private sector input supply, processing and marketing agencies; those growing food crops are drawing on a much wider range of information supply channels and institutions than government alone, and where they do interact with government, they frequently do so not individually but via organisations representing their interests.

An important implication is that the view that 'government must provide' through blanket extension services reaching directly to farmers is outmoded: the most efficient extension services of the future will focus on spheres (geographical; thematic) inadequately serviced by the private commercial sector, which are likely to include soil and water conservation, other environmental, health and safety issues, and the provision of advice (and inputs) to food crop production especially in remote areas. Nor will extension services of the future restrict themselves to direct interaction with farmers: they will also aim to service in different ways and at different levels a multiplicity of intermediate organisations which themselves represent or work with farmers.

Institutional pluralism and farmer participation are important facilitating conditions for effective extension, and are themselves best served by political and macroeconomic climates which encourage: the formation of local groups; administrative and fiscal decentralisation of government agencies, and procedures to allow rural people a voice in the processes of planning and decision- taking. The enabling environment can be further improved by strengthening physical (roads; telecommunications) and social (literacy, numeracy) infrastructure.

Mechanisms for feedback from farmers through extensionists to researchers have been weak in the past and a high proportion of technologies deriving from research have remained non-adopted. Extension structured around a strong management system (T and V) has failed to improve this. The main constraints may lie elsewhere such as in researchers' reward systems, which in many countries are, based more on papers published than on levels of adoption. Similarly, systems for accountability focus more on regular report-writing and financial integrity than on adoption. Both sets of pressures discourage the search for feedback but institutions such as NGOs and farmers' organisations are often better motivated than extension services to exert pressure on research. Donors and governments could usefully search for ways of enhancing the volume and impact of such pressure.

NGOs have developed a number of farmers participatory approaches relying on such institutions as local groups moulded on 'empowering' principles, with volunteer extensionists representing the interests of the group. These experiences have demonstrated the substantial improvements in type and pace of technical change that participatory approaches can bring. However, these approaches are time-consuming and it has not been demonstrated how they might be replicated on a large scale. A shrewd assessment will be needed to screen out unjustified assumptions about the representative ness of groups, the sustainability of volunteer extensionists and the amount of time that poor people have available for meetings, discussions and the like.

It is, therefore, necessary that we continue having appropriate policy interventions in future as well, If we have to attain the growth rate target of 4 per cent in agriculture sector. As such, following areas would need specific attention during the XI Plan period:

- Enhanced capital investment in agriculture;
- Creation of enabling environment to link farmers to markets;
- Credit availability to farmers at low interest rate;
- Announcement of min support pricess well in advance for essential and strategic crops/commodities;
- In view of globalisation of agriculture, a new policy on agricultural exports to capitalize on our comparative advantages –linked with well-organized action plan;
- Major incentives for greater use of growth linked agricultural, inputs (seeds, fertilizers (both mineral and bio-fertilizers), pesticides (both chemical and bio-pesticides), farm machinery and equipment etc.);
- Continued support for buffer stocking of essential commodities at a threshold level (ex: around 15 million tons for cereals) – with creation of ultra modern silos and cold storage facilities to avoid any post harvest losses;
- Insurance of crops and livestock with premium rates that are affordable by the resource poor farmers;
- Incentive oriented and simplified laws and procedures for the establishment of small-scale cooperatives for processing and value addition of their farm produce in rural areas;
- Accelerated pace for the consolidation of land holdings in states lagging behind so far and future land use planning on the scientific and eco-regional basis.

Strategies for Developing Sustainable Approaches and Models

The three vital strategic components based on which the Indian agriculture was meticulously transformed during the planned era included technology packages, institutional support and policy interventions. The radical metamorphosis of agriculture could be made possible through technically qualified manpower developed within the country. The institutional network for agricultural education developed after independence has assumed the status of "National Education System", which is one of the largest in the world. The task of agricultural education in the country is mainly taken care of by 261 colleges. The 42 SAUs and three Deemed-to-be Universities are also meant to impart education in Veterinary, Fishery, Dairying and Animal Sciences.

The research and development efforts targeted for achieving self-reliance in agriculture opened up a series of avenues during this period. The agricultural education system was well set to prepare the graduates to meet the requirements of various public sector organizations. The goal of accelerated agriculture production was largely entrusted to government sponsored public sector departments and institutions. Tailoring the research and development programmes in order to cater to the standards of domestic and export markets through establishment of laboratory facilities for quality control in State Agricultural Universities and Development Departments could not meet the requirements, therefore, special attention required in view of the commercialization of agriculture efforts required to encourage co-operative/ contract farming with buy- back arrangements to initiates market-led production. Implementing suitable programmes for promotion of value added products out of arid zone covering suitable dry land crops.

Promoting the use of eco-friendly inputs, practice of precision agriculture and participatory technology development for sustainable increase in productivity and production of exportable surplus by brining down the cost of production and improving the competitiveness of Indian Agriculture. For exploitation of the agro-ecological diversity of the country, programmes should be evolved for production/ export of agricultural produce from different regions during the off seasons in importing countries. The APEDA and SAD's can work on this and the State Departments of Agriculture, Horticulture, Agricultural Marketing and State Agricultural Marketing Boards can implement this.

Sensitizing the policy makers on creation of community level infrastructure for grading, packing and packaging, storage and promoting grower's association to empower them to negotiate on price. By undertaking the human resource development activities on latest technologies of value addition and post harvest processing capable entrepreneurship at village level need to be developed effectively.

Credit availability must be enhanced by improving Institutional lending to convert agriculture in to agri-business besides creating awareness to use crop insurance programmes and encouraging public private partnership. Adoption of suitable national strategies on augmenting animal productivity, production, consumption and marketing of animal products will generate capital formation per unit of animal resources. Creation of realistic database on meat, milk and egg production, handling transportation, pre and post harvest loss, marketing etc for better economic return. Cutting edge technologies on processing, preservation and value addition on animal products by integrating newer bio-technological tools with the ITK need to be developed. To improve the export oriented meat production, buffalo male calves management with high input ration be made for increased growth rate for marketing. Scientific infrastructure support for in poultry sector, public private partnership may be encouraged in areas like cold chain, processing equipments, distribution net works etc. to produce locally acceptable, cost effective value added poultry products with due consideration of the post harvest losses and efficient utilization of by-products.

Technology specific extension models based on agro-ecosystem need to be developed. Demand driven approach is to be emphasised instead of supply driven approach. Quality extension services like ISO certification, HACCUP, EUREPGAP for both public and private service providers are to be encouraged. Role delineation and linking ATMAs and KVKs for effective implementation of extension programmes at district level is another critical measure to be taken up for proper coordination and integration of extension approaches of Department of Agriculture and Cooperation (DAC), Government of India and Indian Council of Agricultural Research (ICAR). Feedback should be strengthened and its effective use for improving the quality of extension services at the grass root level must be ensured. Participation of front line extension scientist/subject matter specialist in technology generation, assessment, refinement and transfer can play a greater role. Cost economics of technologies are to focussed, accountability of the extension service providers are to be ensured through proper monitoring and involvement of farmers and other stakeholders. Village knowledge centres and Agri-clinics may be linked and integrated with public extension system (SAU, ICAR, line departments) for providing farm advisory services and quality inputs as well. The effectiveness of front line extension of SAUs and ICAR and extension services of state line departments has gone down due to lack of adequate qualified manpower. Vacant posts in SAUs, ICAR and the line departments at all levels should be filled up regularly with priorities.

Capacity building of the staff involved in extension needs to be enhanced for knowledge and management. A functional and organisational relationship between development departments involved in providing extension services needs to be delineated. Village extension committees should be formed. Filtration of farm information of private agencies is required at KVK/ZRS/ATMA to check the misleading information flow to the farmers by different private agencies. Farming system model should be developed for small holders keeping in view the food security at priority and regular cash flow for household needs. Focus of extension programmes need to be given on rural artisans and landless farmers in order to provide them agro-based economic enterprises. As Zonal Research Stations have attained the status of non-performing, therefore, the support to extension as regards to local technological demands is lacking. Therefore, reviving Zonal Research Stations having strong interface with KVKs may enhance the effectiveness of extension system. The ICAR institutions are required to involve and participate more in the activities of KVKs and development departments so that better utilization of technologies developed by these institutions may be done. All institutions such as KVKs, ZRSs, ICAR institutions, development departments, etc. are required to be brought at one platform for agro-climatic zone level in order to chalk out a comprehensive strategy for carrying out technology assessment and refinement requirement of a particular zone and accordingly carrying out various programmes avoiding duplication and using strength of all the available resources.

Dealers of different agencies, who are also engaged in advising farmers while selling their products, are required to be considered as extension agents and trained them accordingly. Contract farming needs to be grounded in a organised way and the state governments are required to bring out effective legislation and enforce law for contract farming. Strong Farmers' Organizations capable of providing extension and advisory services to the farmers, which may be an alternative in pursuing an extension mechanism sustainable on its own. Agricultural credit has to be made available to the farmers adapting simple procedure and cases of suicides need to be seen very specifically. Community radio stations, farm broadcast on low frequency FM band and TV channels are to be used by extension agencies operating at zonal or district level. There is a need for assessing requirement of seed of a district and the concerned agencies should plan and produce required quantity of seed in collaboration with private agencies in terms of using the structures crated by private agencies for flow of advice to poor farmers.

Technological revolutions, climate and environment changes, globalisation of trade, change in economic policies, regional economic integration, developments in information and communication technologies have opened new vistas which have to prudently pursued to accelerate agricultural development process. Growth in agriculture sector would have wider impact in both rural and urban areas. Agriculture has been the largest employer of workforce and its contribution for alleviating poverty in India has been well documented by various studies. Through its strong backward and forward linkages with the other sectors, agricultural growth would have a greater multiplier effect.

This is finding out from the market, what product or products are wanted and in what form *i.e.* customer wants. At this stage one need to know who currently supplies the market? At what times? At what prices? What volumes are sold? How the produce are packed and presented? All these market information will help the extension functionary in understanding either to increase/start supply to the market and answering all these questions is again the process of information gathering. Opinions of knowledgeable individuals, who are commercially involved with trading valuable statistics, such as price data/information on the volumes of produce delivered to the market and the self experience of the extension officer helps in getting fruitful information.

It involves building up an understanding of how produce is distributed and sold and binds the relationships between different sales points in the marketing chain. In this stage, the extension worker needs to understand the various aspects of:

1. Produce distribution system
2. How the marketing system works
3. Marketing margins at various levels
4. Whole sales and middlemen selection as trade partners
5. Information Service available

The following four potential activities can help the extension functionary to achieve the desired goals in marketing extension would be:

- Giving advice to an individual farmer
- Providing market advice to farmer groups particularly through mass extension methods
- Providing advice/information to critical individuals, organisations or private sector companies in the marketing chain whose actions can have a beneficial effect on marketing.
- Adopting a project approach by coordinating the activities of a number of different intermediaries in a marketing chain

With this background, it is imperative to design an effective and efficient agricultural extension model that would enable farmers and others stakeholders to orient their action to the needs of the market and thereby increase the profitability of their operations.

Chapter 25

Innovative IARI Post Office Linkage Extension Model for Reaching the Unreached

R. Roy Burman, J.P. Sharma, S.K. Dubey, K. Vijayaragavan and H.S. Gupta

Background

Indian farming is increasingly facing the challenges and uncertainties due to biotic and abiotic stress, climate change, volatile market, fluctuation of market price. Seed is the most critical input in increasing crop productivity and it was observed that quality seeds of improved crop varieties alone contribute 10-15 per cent increase in yield. Availability of quality seeds of improved varieties is a major constraint. Augmenting farm income is as important as increasing the crop productivity. To address these challenges and increase farm profitability farmers need information related to improved crop husbandry, weather advisory and market information. Technology generation is very fast but extension system fails to deliver the technologies to the end users at the same pace. The ratio of extension worker to the farmer is very wide (1 extension worker: more than 2000 farmers) compared to several developing countries. Agriculture is a state subject and state development departments are responsible for addressing the information need of the farmers. Public sector extension staff of the Department of Agriculture (DoA) of the states dominate extension provision in India. Majority of the states have their staff up to block level. Most of the existing extension personnel perform multiple roles, their visits to the field are irregular and less apart from their pre-occupation with the implementation of subsidy based government schemes. In the above context, it is essential to look for alternative extension approach to reach the farmers with improved technologies to enhance farm productivity and income.

The postal department in India has world's largest network that covers the entire country with its 1,55,015 post office branches. Almost 90 per cent of these post offices are in rural areas (1,39,144 post offices). The rural branch post office caters to 5-15 villages and the branch post master who is the sole staff at such branch post office belongs to the same cluster of villages or its neighbourhood. The village post masters mostly are farmers and they know the local farming situation and farmers' plight well. The trend analysis of post office works showed that during last 10 years, there was sharp decline (about 50 per cent) in the mail and delivery of ordinary post. This has happened mostly after the accessibility and affordability of mobile phone by the rural people and hence, the use of postal communication system reduced to a great extent. The postal department is also interested in diversified activities in partnership with other public and private organisations. The collaborative activities with other agencies like bank, investment agency, insurance departments, etc. had increased (15-20 per cent). This particular finding helped to conclude the possibility of establishing and sustaining the linkage with post offices and utilizing nearly 2.5 lakhs Gramin Dak Sewaks posted at village level in disseminating technology to the farmers. Most of the staff at village level generally belong to the same area and also involved in farming. They know the plight of the farmers and their need very well.

Possibility of Linkages with Department of Post

The analysis of organizational structure, staffing, work load and profile of the post office workers was done to study the possibility of such linkages. A research study conducted in the Sitapur district of Uttar Pradesh revealed that there were 370 branch post offices (BPOs) at the village level manned by 700 Grameen Dak Sevaks (GDS); which should have been 1121, as each BPO is expected to have at least one each of Branch Postmaster (BPM–GDS), Postman and Runner.

The general profile of GDS/BPM indicates that they are mainly the rural farmers living in the same village and discharging the role of GDS/BPM as part-time public sector workers and availing partial benefit from the Department. This indicates the possibility of their inclusion in the additional work of farm technology dissemination in the nearby villages of their operation. The work load of these GDS showed that each of them had to cover on an average of 6–10 villages, 1200 households around the periphery of 8–10 sq. km. It was also found that the average distance travelled by each GDS was 10–12 km/day with a maximum distance of 12 km and a minimum of 1 km/day (Dubey *et al.*, 2014).

With this background, an action research was conceived on establishing linkages with post office during *rabi* 2009-10 by the Division of Agricultural Extension, Indian Agricultural Research Institute, New Delhi for dissemination of agricultural technologies and information to the farmers who are remotely located.

Initiative Taken

In the first phase of the study, the post offices and their personnel at district, block and village level in Sitapur district of Uttar Pradesh were contacted to expiore

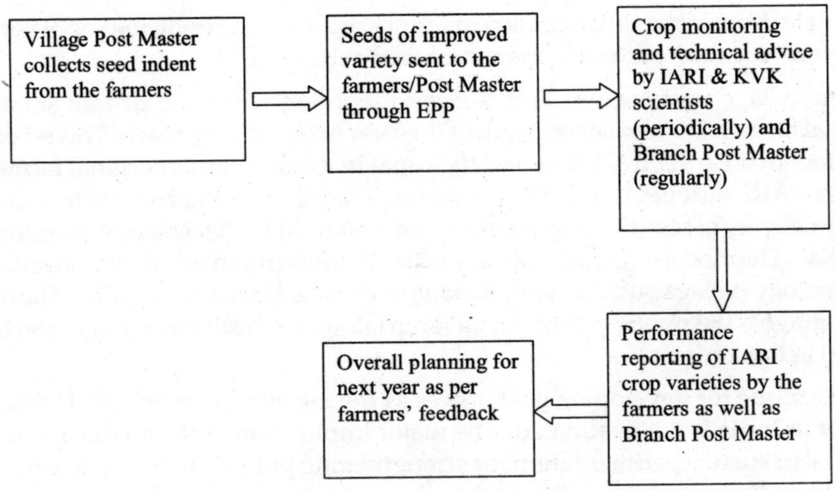

Figure 25.2: IARI-Post Office Linkage Process.

cent of the farmers received the seed of above crops sent through post office within 4 – 6 days of despatch. Yield of major cereals, oilseeds and vegetables increased by 11 – 30 per cent . Dissemination of seeds through the post offices was perceived as effective by the farmers. Majority of the farmers (73 per cent) were satisfied with this new extension model in delivering agro-technology and information services.

Majority of the farmers and village post masters (45.5 per cent) perceived that IARI-Post office linked extension model impacted very highly to increase awareness about IARI varieties. More than half of the farmers (56 per cent) reported that introduction of new variety of major crops had high impact in their area as it changed the seed replacement rate of the area and also the crop diversity of the area. More than three-fourth of the farmers (84 per cent) respondents felt that prestige and social recognition of the farmers in the community increased significantly. This made them "Social Star" in the community. The prestige and recognition of IARI variety growers also went up due to more quality yield of IARI varieties.

Under the cultural impacts, almost half of the respondents reported that the food habit changed in their area as farmers are now growing fine grained IARI paddy varieties. Their children and family members liked the scented fine grain rice more and in special occasion or festival now they were making dishes from such rice. Earlier, only rich farmers could afford aromatic fine grained rice in festival. Another impact which highly affected the community was change in the communication pattern.

Now, frequency of farmers visiting post office and post master increased significantly and post master were acting as opinion leader in the community. Moreover, farmers were not only getting seed but also became aware of different micro saving and insurance scheme like recurrent deposit, life insurance, pension scheme etc. from informal discussion in the post office. As a result, the social network and communication pattern in the rural society changed.

The village post masters and farmers identified "covering distant small and marginal farmers" as a major strength of this new extension approach. This is because of the fact that earlier it was beyond the capacity of small and marginal farmers to avail the IARI varieties due to long distance. The other strength of the model were "timely delivery of seeds", "cost effectiveness" followed by "technology backstopping by KVKs". The model emerged as an alternate extension mechanism for dissemination of technology (village post masters as community based extension agents). The model also highlights the public-public partnership (Research institution and Department of Post) in this process.

The scope for developing post offices as the means of agricultural technology transfer in India is now validated. The major implication of the model affirms the potential of such experimentation for strengthening public-public linkage for farm technology dissemination which can later be institutionalized as the effective model of front line transfer of technology by the large number of research institutions. The strong network of quality human resource manpower of village level postmasters shall definitely complement and supplement the existing public sector extension personnel and ultimately help in reducing the extension worker: farmer ratio in the country.

Future Plan of Work

To institutionalize the model in wider area, the model has been expanded further in partnership with the KVKs. The Krishi Vigyan Kendra (KVK) at the grassroots level is our partner. Their role is to facilitate and provide technology backstopping to the village post office staff and the farmers and identify suitable technological interventions, crops and varieties in a particular location. Role of stakeholders including research institutions, KVKs and postal staff and their linkage mechanism in multi partnership model needs to be formalized. Capacity building of stakeholders through demonstration, training and 'media-post' needs to be strengthened. Though the 'Media post' scheme of postal department, posters on improved varieties and crop management practices will be displayed at the branch post office.

The KVKs are expected to identify the branch post masters in the remote areas of the district and sensitize them about the approach and the role of branch post masters in this process. Quality seeds of improved crop varieties are being sent to the KVKs through post. Demonstrations of improved varieties are laid out in KVK farm and in the field of branch post master and the identified farmers. Seeds multiplied in the KVK farm will be sent to the farmers through post office in the next season. Future strategy for up-scaling this model will involve formation of farmers association for multiplication of quality seeds of improved IARI crop varieties under the technical supervision of the KVK scientists and selling the quality seeds through retail post scheme of postal department. This will ensure availability and self-sufficiency of quality seeds of high yielding crop varieties at the grassroots level.

Expected Outcome

The major expected outcome of this innovative extension model is reaching remotely located farmers with technologies and information through post office and village post masters (who will act as village agents) which will increase farm productivity and income. The model will be one of the cost effective extension models. This innovative approach may be named as 'Postal-Agri Extension Services'.

Chapter 26

Strategies for Effective Transfer of Technology for Increased Sugarcane Productivity

T. Rajula Shanthy, Jitendra Chauhan and A.K. Singh

Indian agriculture had been on traditional lines till the first waves of Green Revolution in the late 60s, which gave a sudden boost to the production and productivity of major crops. Quick dissemination of information from the Agricultural Research System to the farmers and reporting of farmer's feedback to the research system is one of the critical inputs in transfer of agricultural technology. As a result of extensive research over the last several years, technologies are now available to boost sugarcane production in the country. However, the impact of such massive research efforts could not be fully reflected in terms of cane productivity. In most of the developing countries today, it is not the lack of technology that worries, but it is the rate of transfer of technology from the points of production (research system) to the points of utilization that matters.

Sugarcane Scenario

Our country has come a long way in sugarcane cultivation from just 1.19 m. ha area under sugarcane in 1930-31 and a sugarcane production of 36.35 m. tons to 5.15 m. ha area with a record production of 375 m. tons sugarcane during 2013-14. Correspondingly, the productivity has also increased from 30.9 t/ha to 69.0 t/ha during 2013-14. However, there is a wide disparity in productivity between the different cane growing states of the country. This gap has to be bridged so as to maximize the unit productivity and minimize the cost of sugarcane cultivation.

The Indian sugar industry lacks the flexibility of the Brazilian sugar industry, where the processing of cane for sugar or alcohol depends on the prevailing market trends with respect to these commodities, which ensures sustained growth and profitability of the industry. Domestic sugar price of sugar in India is among the lowest in the world. The production cost of Indian sugar is estimated to be in the medium range - costlier than Australia and Brazil but lower than that of USA. In future, we may face stiff competition from African countries, whose production costs are lower than India (Nair, 2014).

Sugarcane growers in India seem to have their own valid reasons for non-adoption of sugarcane technologies, which could be ignored altogether in 'top-down' approaches. The list includes

- ✰ Technology is too complex
- ✰ Not easily divisible into manageable parts
- ✰ Not compatible with farm and personal objectives
- ✰ Not flexible enough
- ✰ Not profitable
- ✰ Too high capital outlay
- ✰ Too much additional learning is required
- ✰ Risk and uncertainty is too great
- ✰ There is conflicting information
- ✰ Lacking the physical infrastructure and
- ✰ Lack of a social infrastructure

Technology transfer typically involves a 'top down' approach where scientists determine research priorities; generate innovations they believe are good for the farmers and provide results to extension agents. This is sometimes termed the linear adoption or diffusion model (Rogers, 1983). Extension efforts that take farmers into account need to be strengthened and this should form one of the primary tasks towards achieving the sugarcane targets for 2025 AD and beyond.

A Systems Perspective

The usefulness of a systems approach for understanding and analyzing agricultural technology generation and dissemination is widely acknowledged. Although they are sometimes criticized because they are so abstract, systems analyses offer holistic vantage points for understanding the factors that impede or enhance the two-way flow of technology and information between farmers and the public organizations that constitute the system.

Any agricultural knowledge system consists of four components set in a larger context (Figure 26.1). The components are technology generation, technology transfer (knowledge and input transfer), technology utilization, and support system. The organizations that constitute the components, as well as others in the system environment influence each other in complex ways.

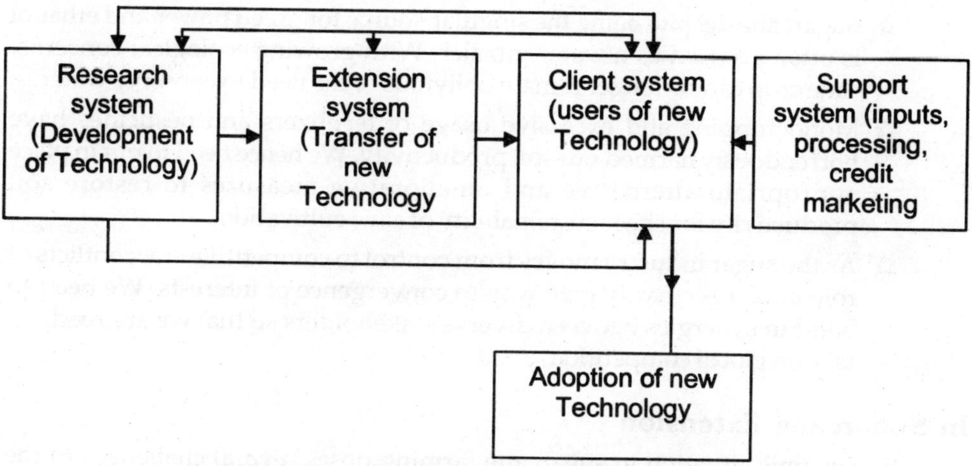

Figure 26.1: Agricultural Knowledge System.

Emphasis on a systems perspective in agricultural research requires a complete reversal of the technology generation approach *viz.*, a farmer first or farmer participation approach where farmers are fully involved at all stages of technology development (Chambers, 1983, Van den Ban and Hawkins, 1998). Participatory research is a systematic process involving people in the analysis of their own situations. Farmer participatory research is a term increasingly used in agricultural research and development circles. In the context of shrinking resource base, agriculture in future can be sustained only by producing more food, fuelwood, fibre and other commodities from less land, less water and less energy per unit of output without the accompaniment of ecological harm. The stress on 'farmer participation' has arisen from a realization that earlier approaches to on-farm research recognized the importance of farmers, but was not able to effectively incorporate farmers' skills and experimental practices into the research process.

The Strategic Extension Campaign (SEC) methodology developed by Food and Agriculture Organization also emphasizes the importance of people's participation in strategic planning, systematic management and field implementation of agricultural extension training programmes.

Challenges Faced

In Sugarcane Agriculture

Sugarcane crop currently faces severe challenges on several fronts in our country. This includes:

- Increasing cane production to meet the ever increasing demand for sugar
- Diminishing interest in agriculture due to decreasing net return from the crops
- Diversion of labour from agriculture to other sectors. The fact here remains that adequate farm labour is unavailable and the available labour is unaffordable

- ☆ Sugarcane despite being the singular source for green power and ethanol is often accused as a water guzzler. With growing ecological concerns, cane crop can no longer remain oblivious of the need to conserve water.
- ☆ Monocropping and excessive usage of fertilizers and pesticides have horrendously harmed our soil productivity. We hence need to put in place appropriate alternative and ameliorative measures to restore soil productivity for sheer sustainability of cane cultivation.
- ☆ As the sugar industry moves from control to competitive era, conflicts of role must necessarily give way to convergence of interests. We need to build in synergies between diverse stakeholders so that we are ready to take on global competition.

In Sugarcane Extension

The dynamic situation in sugarcane farming poses several challenges to the extension system. Some of the challenges with which the extension system has to cope up include:

- ☆ How to reach individual farm families involved in the sugarcane cultivation and deal with the Complex, Diverse and Risk prone (CDR) environment?
- ☆ How to improve the technology standards of the farmers through effective and optimal technology use, especially when the pressure on land is increasing and common property lands are slowly fading out.
- ☆ How to face the emerging agricultural development situation as a sequel to the technological and development interventions?
- ☆ How to take cognizance of the changes that are taking place in the society in terms of Shift from farming to industry, rural to urban (migration), focus from social to economic issues.

Way Forward

Sugarcane agriculture is faced with several challenges, as also uncommon opportunities. Being a crop of a well-organized sector, namely the sugar factories, public private partnership as a governance strategy can very well be utilized to converge the efforts of varied organizations at the grass root level of a community. The sugar industry contributes substantially to the rural economy as the sugar mills are located in rural areas and provides large scale employment to rural population. A SWOT analysis of the extension approaches till late 1980s reveal a lacuna of trickle down approach rather than a democratic one. Change thus became inevitable in all developmental spheres of steering India into a developed country. In this era of globalization, mankind has become assertive and needs a rationalization of any developmental efforts being undertaken. This can include:

Technology Generation

A critical technology is a technology aimed at solving of a current problem faced by users in their context, and a technology should be developed in terms of the capabilities, resources, opportunities, social and economic considerations that are both internal and external to the technology with emphasis on positive and negative consequences for the users (Gupta, 2009). Involvement of farmers in the planning, operating, monitoring, evaluation and follow-up actions in technology generation and dissemination is of tremendous importance. Nevertheless, the attributes of technology *viz.*, relative advantage, compatibility, complexity, trialability and observability have to be kept in mind when making specific recommendations.

Public Private Partnership (PPP)

PPP through participatory mode has been recognized as one of the most effective mechanisms to upgrade the country's infrastructure services. There is a scope to leverage PPPs as a relevant means in the agriculture sector as well. Enhanced yield and productivity is a crucial need, with India still battling food insecurity and poverty. Participatory approaches in agriculture are generally found at the community level where the strengths of the public and private sectors supplement each other in providing information and advisory services that address the needs of farmers and rural communities. The public sector's obligation for provision of information and services can be best achieved through harnessing the potential of the private sector to add local context in a commercial environment.

Convergence of Extension Services

There are many extension service providers in the field, providing different kinds of useful services like information and service support to farmers. They are state, central government agencies, agribusiness companies, agri-preneurs, input dealers, manufacturing firms, NGOs, farmers' organizations and progressive farmers. There is duplication of efforts with multiplicity of agents attending extension work without convergence. There should be a coordinated attempt to synergize and converge these efforts at the district level through ATMA and below to improve the performance of various stake holders.

Increased Role of Extension Agents

In addition to the government efforts in stepping up the allocations for sugarcane development programs, the sugar factories / department of agriculture through their extension agents are expected to play an increasing role in improving the productivity levels in their respective zones. Such programs would enable the farmers in each zone to improve their sugarcane yields simultaneously enabling the sugar factories to get larger quantities of better quality cane at appropriate time, thus ensuring higher sugar recovery levels with a longer crushing duration. The extension agents can use the following ways to help the cane growers to reach their goals:

- Giving timely advice to make them aware of a problem
- Increasing the range of alternatives from which they can make a choice
- Informing them about the expected consequences of each alternative
- Helping them to decide which goal is most important
- Helping them to make decisions in a systematic way, either as individuals or as a group

- Helping them to learn from experience and from experiments and
- Stimulating them to exchange information with colleagues

Frontline Demonstrations

The adage 'Seeing is believing' holds good even today. Realizing this, the Indian Council of Agricultural Research introduced the concept of 'Frontline demonstrations'. These are first line demonstrations conducted by the research scientists to demonstrate promising technologies in the farmers' fields under different agro-climatic regions and farming situations. The purpose is to convince extension functionaries and farmers together about the potentialities of technologies for further wide scale diffusion. These demonstrations are conducted in a block of at least one hectare land in order to have better impact of the demonstrated technologies. The special features of frontline demonstrations are:

- These are conducted under the close supervision of the scientists.
- Only newly released technologies or those likely to be released in near future are selected.
- Only critical inputs and training are provided from the scheme budget, remaining inputs are borne by the farmers themselves.
- Training of the farmers associated with the frontline demonstration is a pre-requisite for conducting such demonstrations.

Sugarcane Breeding Institute, Coimbatore has been conducting frontline demonstrations since 2001-02 in Coimbatore and Tirupur districts in Tamil Nadu state. There was considerable improvement in yield in all the demonstration plots over the farmers' practice. Since the demonstrations are conducted in farmers' fields they serve as a motivation for the farmers in the neighbourhood as well. While demonstrating the technologies in the farmers' fields, the factors contributing to higher crop production and field constraints of production were also studied to get feedback information and discussed in scientific meetings (RajulaShanthy, 2011).

Pluralistic Extension

Of late, development practitioners have termed an "Integrated, Multi-Disciplinary and Holistic approach to Rural Development (IMDH-RD)", stressing the need to address both agricultural and rural problems to create effective and environmentally sustainable development (Davidson, 2006). To effectively cope with the challenges posed by a multi-sectoral approach, a pluralistic agricultural extension mechanism is required to coordinate the various stakeholders. Pluralism recognizes both the heterogeneity of the farming community and thus the need for diversity in extension service delivery systems. This approach requires farmers and other rural inhabitants, extension agents, input supply dealers, researchers, policy makers, and so on to focus their efforts, activities, and programs on encouraging a sustainable agriculture that is inclusive rather than exclusive.

communication technologies will facilitate training and information exchange. Public extension's future (such as the state departments of agriculture) depends on our ability to interpret trends and use information technology to deliver programs and teach problem solving for the farmers. But such strategy will need to be supplemented by other extension strategies so as to enhance sugarcane productivity and ensure increased profits for the cane growers.

Conclusion

The conclusion is that no single theory, model or strategy is likely to be sufficient by itself. If the possibility of adoption is to be maximized, communication channels must be utilized in an ideal sequence, progressing from mass media to interpersonal channels (Sill, 1958). The production of sugarcane has an impact in rural areas and has served as an effective instrument for carrying progressive trends in the countryside. The development of sugarcane on a holistic sense in the years to come requires mix up of innovative extension approach with technological interventions appropriate for a given situation. It is just the fitness of such blending that demands to formulate and implement a strategic plan for total sugarcane development.

References

Chambers R (1983) Rural Development: Putting the last first, Longman, Essex.

Davidson AP (2006) Enhancement of extension systems in agriculture. Asian Productivity Organization. Tokyo.

Gupta DD (2009) Sugarcane development – Technological interface between tradition and modernity. Agrobios, (India), Jodhpur pp. 359.

Nair, NV (2014) An overview of sugarcane agriculture in India and the future scenario. In: T Rajula Shanthy et al (Eds.) Recent technologies for increase sugarcane productivity. ISBN 9789380800257. PRDAG Print, Coimbatore. p.1-7.

Rajula Shanthy, T (2011) Strategies for effective dissemination of appropriate technologies to sugarcane growers in India. SugarTech 13 (4): 354-359.

Rogers EM (1983) Diffusion of innovations. Free Press, New York.

Sill M (1958) Personal, situational and communicational factors associated with farm practice adoption, PhD thesis, University Park: Pennsylvania state University.

Van den Ban AW and Hawkins HS (1998) Agricultural Extension. Oxford: Blackwell.

Index

B

C

D

E